A Cultural History of Reforming Math for All

Although many accept that math is a universal, culturally indifferent subject in school, this book demonstrates that this is anything but true. Building off of a historically conscious understanding of school reform, Diaz makes the case that the language of mathematics, and the symbols through which it is communicated, is not merely about the alleged cultural indifference of mathematical thinking; rather, mathematics in schools relates to historical, cultural, political, and social understandings of equality that order who the child is and should be. Focusing on elementary *math for all* education reforms in America since the mid-20th century, Diaz offers an alternative way of thinking about the subject that recognizes the historical making of contemporary notions of inequality and difference.

Jennifer D. Diaz is Assistant Professor of Education at Augsburg University, United States.

Routledge Cultural Studies in Knowledge, Curriculum, and Education

Series editors: Thomas S. Popkewitz and Daniel Tröhler

A Cultural History of Reforming Math for All

The Paradox of Making In/Equality

Jennifer D. Diaz

Foreword by Thomas S. Popkewitz

Routledge
Taylor & Francis Group

LONDON AND NEW YORK

First published 2018 by Routledge

2 Park Square, Milton Park, Abingdon, Oxfordshire OX14 4RN
52 Vanderbilt Avenue, New York, NY 10017

Routledge is an imprint of the Taylor & Francis Group, an informa business

First issued in paperback 2019

Library of Congress Cataloguing-in-Publication Data
Names: Diaz, Jennifer, author.
Title: A cultural history of reforming math for all : the paradox of
    making in/equality / by Jennifer D. Diaz.
Description: New York : Routledge, 2018. | Series: Routledge
    cultural studies in knowledge, curriculum, and education ; 4 |
    Includes bibliographical references.
Identifiers: LCCN 2017017486 | ISBN 9781138638402 (hardback)
Subjects: LCSH: Mathematics—Study and teaching—United States. |
    Educational equalization—United States. | Curriculum change—
    United States.
Classification: LCC QA13 .D525 2018 | DDC 510.71/073—dc23
LC record available at https://lccn.loc.gov/2017017486

ISBN: 978-1-138-63840-2 (hbk)
ISBN: 978-0-367-87712-5 (pbk)

Typeset in Sabon
by Apex CoVantage, LLC

For Jesse and Ada, my *all*.

Contents

Series Foreword

Routledge Cultural Studies in Knowledge, Curriculum, and Education

This series brings interdisciplinary studies that focus on knowledge of education, school, school reforms, curriculum, and research projects. Central to the studies are how "we" think, order and act in schooling—its systems of reason or languages—as cultural practices about how life is to be lived and how the possibilities of the future are envisaged. The studies are referred to as "critical" in the sense that they make visible the principles governing and regulating both, what is known, done, and acted on in schooling and in research.

The series provides ways of re-conceptualizing approaches to the study of school change and reform in policy, curriculum, and teacher education. It takes what are given as natural and taken-for-granted in the everyday life of schooling to ask about the conditions that make possible its objects of reflection and action. The series is to provoke thoughts about schooling as cultural practices to think about the principles that order the construction of the subjects and subjectivities of schooling.

Jennifer Diaz's The Paradox of Making In/equality: *A Cultural History of Reforming Math for All,* the fourth volume in our series, contributes to this purpose through its study of mathematics education. Mathematics has become one of the privileged knowledge in contemporary societies, traversing the fields of science, technology and engineering associated with the new economies. It is thought of as central to economic success of the nation associated with the new global context. Mathematics is also thought about as a "tool" necessary for the social competence of the citizen in a world where statistical knowledge, science and state efforts coincide for producing reforms and organizing social change.

As previous books in the series, it makes visible the paradox of contemporary efforts to recognize differences that inscribes differences. The book explores how the very concrete strategies to include children embodies its own paradox—practices that exclude and abject in the efforts to include. The focus of this study is how what seems as a neutral mode of knowledge-mathematics- when assembled and connected in pedagogical programs, embodies unspoken but present principles about kinds of people who are different, such as the disadvantaged, the at-risk, and "the child left behind. It considers modem pedagogy as historically related to the art of governing the modern state. In this sense, the knowledge of schooling—its system of reason or language—functions as an "actor" in the making of children and the production of difference. Through the historical study of mathematics education reforms in the US, the study makes visible how the very knowledge or system of reason that orders and classifies the practices to correct social wrongs reinscribe the very principles of difference and exclusion they are to challenge.

The argument of the book brings into focus the curriculum of school subjects. Typically taken-for-granted in studies of education, Diaz explores the translation of mathematics into mathematics education to historically examine the changing cultural principles about "the nature" of the child that orders and classified what counts as the "mathematically literate" child. In doing so, she explore that alchemy of the school subjects, the magic translation and transformation of disciplinary knowledge into texts about what children are to know and how that knowing is to be obtained. Embodied in this argument is that schools require alchemic processes as children are not scientists or mathematicians. Thus, translations are necessary. But the question that Diaz historically engages in is the translation models that organize the school curriculum. The translation tools, the book persuasively argues, have little to do with understanding disciplinary knowledge. The models of curriculum emerge from psychologies of the child that are moral sciences that inscribe principles of difference.

Diaz's book challenges the very conventions and traditions of educational studies that underlie current efforts to understand the effects of schooling and its efforts for school reform. She argues that what is taught as the knowledge of an academic field, mathematics, are translation practices concerned with making kinds of people that produce differences and divisions. The argument views mathematics education as an alchemic process in which cultural theses are generated about social exclusion and abjection in the trust to include. Through analyzing how the curriculum embodies principles about kinds of people since the reform movements since the Post-World World Two United States, the book makes possible an understanding how efforts about "all children learning" mathematics embodies differences about children that divides and racializes through cultural distinctions about the child that have little to do with *mathematics* education.

Diaz's historical study about the paradox of equity/inequity has important implications for understanding the political of schooling; that is, the manner in which and the conditions on which the school curriculum, models of teaching, and the psychologies of learning embodied principles of inequality and cultural differences. The broader implications of the theoretical concerns in the study of differences have implications for understanding issues of school knowledge in different historical spaces and national contexts.

Diaz's book contributes to this series about educational scholarship through theoretically integrating into an historical analysis to explore the limits of the principles that order the "commonsense" practices of schooling.

1. The analysis makes the systems of reason or languages of schooling as cultural practices and central to its study. In particular, the book explores the magical translation and transformation of a field of disciplinary knowledge into content knowledge of mathematics education. The alchemic practices are explored over time to examine the paradox of education as historical practices that change over time. These changes have a continuity. They produce particular frameworks for ordering and classifying people and their difference; classification that have little to do with the naming of the school subject, such as learning of mathematics. The book provides a strategy for re-thinking central principles that order what is done and said in the school curriculum.

2. It brings together different interdisciplinary approaches from the humanities and social sciences concerned with the materiality of "knowledge"—that is, knowledge as construing and constructing who "we" are and should be. In contributing to the series, the book provides ways to rethink, revise, and reframe contemporary school studies built on the categories of representation and identities given as the origins of descriptions and the location of causes in schooling.

3. The series provides alternative ways of thinking about the role of academic research and the social commitments to change. By focusing on the system of reason or language of education, Diaz's study provides a way to think about, and unthink, the assumptions, consequences and implications of the contemporary conventions and distinctions.

Thomas S. Popkewitz,
University of Wisconsin-Madison

Daniel Tröhler,
University of Vienna
3 June 2017

Foreword

Contemporary international and national educational policy, curriculum, and research in reform seem to converge on two twin commitments: to prepare children for the future, sometimes called the knowledge society or knowledge economy, and to provide equal educational opportunities for "all children," a topos circulating in the international student performance assessment as well as in national U.S. policies and programs. The commitments to school improvement have a salvation theme about hope for the future society and, to borrow from Adam Smith, the wealth of nations.

These commitments are connected to changing instructional processes, teacher education, and psychologies of learning through school reforms and research. In one strand of this research, attention is given to "making" the professional teacher though teacher education preparations. Classroom analyses focus on creating the pedagogical expertise for children to learn the content of the school subjects to participate as reasonable citizens in the future. The professional, expert teacher is one who has a common core of knowledge and engages in high-leverage teacher practices to enable children's learning of the curriculum.

Central to this research and reforms are mathematics and science education. The most prominent research program directs attention to micro or core practices of teaching. The research identifies the necessary knowledge, skills, and dispositions that allow a novice classroom teacher to utilize the professional expertise necessary for "successful" teaching. At the macro level of international assessments of children's performances, the Organization of Economic Development and Cooperation's (OECD) Program for International Student Assessment (PISA), for example, measures 15-year-old students' reading, mathematics, and science literacy every three years to assess nations' progress in mathematics, science, and literacy education. The assessments are linked to identifying the practices of teacher recruitment and support as well as personal and social attributes that contribute to an effective school system.

I begin with these foci on teacher and teacher education reform as, first, they privilege science and mathematics education as essential to the health and wealth of nations. Given the acronym of STEM (science, technology, engineering, and mathematics), curriculum and teacher education reforms show no bounds in their global prophecies and salvation themes. STEM education is to make possible the national ability to maintain economic prosperity by enabling future generations to obtain jobs in STEM fields. STEM also fulfills a cultural and political discourse about education, creating a literate citizenry in a world increasingly governed through science, mathematics, and technology. Science and mathematics, it is argued, are necessary for people to understand and navigate the complexities of modern societies. Mathematical modeling, for example, is forecast as enabling children to make better judgments in their life choices and civic responsibilities.

The Paradox of Making In/Equality places the salvation themes of schooling under scrutiny. It examines the rules and standards of the "reason" of mathematics education to understand historically how the qualities, characteristics, and differences thought of as "the nature" of children are inscribed in the school curriculum. Focusing on American mathematics curriculum from postwar reforms, the study makes visible how mathematics education embodies cultural theses about different kinds of people. These cultural theses are embodied at the intersection of the distinctions that order school policies, sociologies, and psychologies of children's learning and the selection and organization of mathematical relations and symbols that children are to learn.

The study begins with a simple, seemingly innocuous property of mathematics education: the equal (=) sign. Making a historically important observation, Diaz argues that the mathematical equal sign (=) is never merely an expression of pure numerical equivalences. The equal sign in the school curriculum becomes part of discourses that connect with political theories of equivalence that carry norms and values about equality and inequalities. These norms and values, as Diaz explores, are inscribed as characteristics of children's learning that change over time in the reforms.

What does not change, however, is the paradox of mathematics education in producing double gestures that appear in tropes about teaching "all children." One gesture is the hope of schooling to create a moral order through making each child into a particular kind of person who productively participates in society as a citizen. Simultaneously with this gesture of hope are principles generated about populations envisioned as dangerous to the future. The social and cultural distinctions circulating in the sociologies and psychologies of the classroom have the duality of defining the mathematically (il)literate child. The divisions create a continuum between normalcy and pathology. These divisions are not about learning mathematics. They are cultural distinctions related to social exclusions and abjections.

Diaz's thoughtful and careful historical analyses dissolve salvation claims that attempt to identify "best practices" and produce the educational system for "all children." It also raises questions about contemporary orthodoxies of educational research. The claims embody a series of chimeras and paradoxes. The chimera, an illusion and fabrication, is that the mathematics being taught to children are translations that expunge the norms of participation, truth, and recognition in the field of mathematics. What is taught as mathematics education is something different. The school subject of mathematics education entails complex sets of relations and principles of recognition about cultural and social differences of kinds of people.

The paradox is in the concrete practices of ordering what is known and how embodied in that knowing are double gestures; the hope of making productive citizens simultaneously generates fears about the dangers and dangerous populations that threaten that hope. The differences in the characteristics of children appear as classifications and distinctions about the child to learn mathematics. The double gestures are embodied in the distinctions that describe the child as one who is/is not motivated, attentive, problem-solving, and so on.

Diaz's argument leads to a (re)thinking and (re)visioning of school reform and educational research through focusing on mathematics education as analogous to alchemies. The arguments about the alchemy begin with a simple and profound observation. Pedagogy is as the medieval metallurgy that sought to transmute base metals into gold. A magical transmutation occurs as academic knowledge is moved into the space of schooling. The translations are acts of creation that are connected and assembled in relation to the cultural principles and social patterns governing schooling. The principles of the alchemy are no longer those of mathematics or science but those historically tying curriculum content to pedagogy.

This interrogation of the changes in the alchemy of mathematics education and its system of reason over time point to a significant element elided in current STEM educational reforms. It is possible to say that whereas the name of the curriculum signifies teaching mathematics and science, something different is practiced. The translation models of school curriculum assembled through principles of classroom management and learning psychologies that were not created to understand the disciplinary practices named as school subjects—mathematics and science, for example. The curriculum principles were formed in relation to social and political projects about making kinds of people, such as the modes of life associated with the citizen, the lifelong learner, and the mathematically literate child in contemporary reforms. And with these fabrications of kinds of people are the qualities of people who are not learners and placed in spaces outside the normalcy inscribed in schooling.

Diaz's study raises significant questions to consider not only in mathematics education but also in how educational enactments of social commitments are inscribed into the systems of reason of pedagogy to produce social exclusions and abjections in efforts to include. This conclusion requires thinking differently about what schools do and the political in education. This is a different way of thinking than what is given in the salvation languages and theories of learning in pedagogy. The political is in the differentiations and divisions historically generated in schooling about who children are and should be. The ordering and classifying practices of the curriculum produce divisions that create cultural spaces of normalcy and pathology. The challenges of the historical analyses of mathematics curriculum makes visible how schools instantiate principles of stability as a strategy of change. Further, salvation themes to prepare for the future exclude and abject in efforts to include.

This study of the systems of reason that order the curriculum and its inscriptions of the child enables the study of schooling to understand the limits of the politics of representation that are not only about the *doxa* of school reform and research. It leaches into critical pedagogy that accepts the models of curriculum to leave unquestioned how differences are made through concrete rules and standards that order reflection and action in the school subjects.

The Political of Schooling: Making Kinds of People

The significance of this book, *The Paradox of Making In/Equality*, is to direct attention to how the organization, selection, and evaluation of subject matter in the modern school are political. That political is not about representing social interests but in what Foucault (1979) has called *governmentality* and the political philosopher Jacque Rancière (2007) discusses as that partition of the sensible and sensibilities. One of the significant "facts" of modernity is that power is exercised less through brute force and more through systems of reason or knowledge that order and classify what is known, seen, talked about, and felt. This concern about the political is expressed in strands of feminist, critical race theories and post-foundational studies. The literature asks how identity, subjectivities, and differences are produced through social and cultural technologies of modernity. The concept of governing refers to the system of reason or maps produced that partition what is "sensible" as the objects seen and acted on in teaching, the child's thought processes, and the "nature" of disciplinary knowledge translated for instruction (see, e.g., Popkewitz, 2008; Popkewitz, Diaz, & Kirchgasler, 2017).

This conception of the political leads to a second observation that emerges from Diaz's study of mathematics education. Schooling connects the scope and aspirations of public powers with the personal and subjective capacities of individuals. Modern governing linked two seeming opposites: the freedom and will of the individual with the political liberty and will of the nation. The American or French Revolutions and subsequent writings saw that the future of the republic, and later notions of democracy, embodied particular cosmopolitan values and dispositions about science, autonomy, and participation that gave the state its reason for being. "Citizens are not born; they are made" (Cruikshank, 1999, p. 3). Democratic participation was "something that had to be solicited, encouraged, guided, and directed" (p. 97). As Wagner (1994) argues, "The history of modernity cannot simply be written in terms of increasing autonomy and democracy, but must be written rather in terms of changing notions of the substantive foundations of a self-realization and of shifting emphases between individualized enablements and public/collective capabilities" (p. xiv). This governing of freedom is one of the ironies of the modern republic, liberal democracy, and pedagogy.

Science and mathematics have played a presumably practical role in the technologies of making the citizen. In the 19th century, science and mathematics were brought into cultural theses about the cosmopolitan reason that linked the making of a citizen, the nation, and progress. Foundation stories were told about Americans transforming the wilderness into "a prosperous and egalitarian" cosmopolitan society whose landscape and people had a transcendent presence through its technological achievements (Nye, 2003, p. 5). The technological marvels of the railroad, electricity, bridges, and skyscrapers were placed in a cultural dialogue about the national manifest destiny (Nye, 1999). The natural power of Niagara Falls, the Grand Canyon, and technologies represented in the railroad, bridges, and city skyscrapers were narrated as a causal chain of events of an inevitable developmental process. The faith in science and mathematics as the apotheosis of human reason was not only in North America. Salvation theses of science crossed the North Atlantic, ranging from the Fabian Society to the German Evangelical Social Congress, the French Musée Social, and the Settlement House movements in many countries (Rodgers, 1998).

The inscriptions of science and mathematics were not only about creating the citizen. It embodied a comparative style of thinking and the problem of exclusion and abjection that earlier, I discussed as double gestures. These double gestures were given expression in the turn of the 20th-century social and educational reforms that evolved around the Social Question that was central to the political, social, and educational programs associated with American Progressivism. The Social Question entailed various political, social, and educational reforms that brought science in to change the moral and economic disorder associated with urban conditions and its populations. The double gestures of the Social Question in the United States expressed the hope of a new social and political order produced through industrialization that simultaneously expressed fears of the loss of moral order.[1] The hope of the new welfare state programs was to eliminate the social evils of the city by active intervention in the conditions and everyday life of the city. The fear was that if childhood development was not controlled, the child would become potentially dangerous to the future of the republic (Krug, 1972). G. Stanley Hall (1928), a founder of child studies, gave importance to the study of adolescence as important to the regulation of the transition of youth to adults, who he said were living in "the urban hothouse."

When mathematics education is looked at in relation to the Social Question, its trajectories and distinctions in the curriculum embody the double gestures of the hope of future citizen and fears of the dangerous people threatening that hope. The teaching of arithmetic in the connection-ist psychology of Edward L. Thorndike, for example, was to rationalize the sequences of teaching mathematics to produce a self-directed and self-governed individual. Thorndike (1912/1962, p. 145) applied a utilitarian notion to teaching to order moral conduct: "Knowledge pertaining to moral conduct is thus above knowledge pertaining to manners; knowledge pertaining to health is about knowledge pertaining to wealth" (Thorndike, 1912/1962, p. 145). Thorndike's studies of mathematics education incorporated eugenics to differentiate the manner of life in which children of different social backgrounds and gender were to use mathematics.

I focus on the Social Question and the inscription of a comparative mode of reasoning to focus on the importance of the historicizing that occurs in this book. With all of the contemporary concerns of the school providing equality and equity, Diaz's analysis explores how the Social Question remains. Urban education is still a signifier to recognize the inclusion of differences that inscribes difference. But the classification of "urban" and the Social Question today have different sciences that order and (re)vision the principles of inclusion and exclusion as reforms move from the 1950s to the present. The comparisons and divisions that are obscured in today's utilitarianism focus on how to make effective and more efficient standards of communication, participation, and social relationships in the classroom. Such research and reforms leave unproblematic the alchemy that translates disciplinary fields into the curriculum and the paradoxes of inclusion and exclusion they embody.

Alchemy

The importance of *The Paradox of Making In/Equality* to the study of schooling can be further examined through its historical focus on schooling in making kinds of people and differences through the cultural analysis of the principles embodied in the curriculum of school subjects. As I mentioned earlier, contemporary policy, curriculum reforms, and research take for granted that schools are teaching science, mathematics, literacy, and so on. This naturalizing of school subjects is found in phrases of policy and research about "the achievement gap" and STEM education. The reform tasks seem to be about finding the best practices and the effective, high-leverage teaching strategies that will solve the problems of quality and inequality. This taking the curriculum models of school subjects as the origin of school change circulates with studies of critical pedagogy. The assumption is that what is taught matches the names given as the school subjects—mathematics, science, literacy, and so on.

What these studies ignore, and what Diaz historically makes visible, is that what is taught has little to do with the cognitive domains that are given to school subjects. They are translations that displace, transform, and change the disciplinary fields of knowledge to other forms, levels of abstraction, and practices of actualization. What is forgotten is that translations are never copies. They are creations.

Diaz brings the notion of alchemy to think about the processes of creating the curriculum, liking mathematics to the sorcerer of the Middle Ages who sought to turn lead into gold. Pedagogy is such a magical transformation that transforms mathematics, science, and music education, for example, into rules and standards connected to educational psychologies about learning, conceptions of childhood, and the administrative activities of the teacher for governing the child.

I make this observation to underscore, at one level, that alchemies are part of the school as children are neither mathematicians nor historians. Translation tools are needed for instruction. Although the labels of the school subjects and academic disciplines signify that children are learning science and mathematics that carry forth the salvations themes of education in making the future, something else is occurring!

But under scrutiny are the particular and concrete ways that the curriculum translates and transforms the curriculum as the order of things and the double gestures that normalize and pathologize. If the formations of school subjects are examined, for example, they embody the double qualities of the Social Question. The narrative structures and ethical messages of literary texts in the British mass schooling of the 19th century, for example, were to produce the moral and physical well-being of children, who embodied the will of the nation and its images of progress through making the stories relevant to the everyday experiences of the "inarticulate and illiterate" of the working classes working-class children (Hunter, 1988). Embodied in the distinctions were fears of the threats to progress and the narratives of moral order. Art and the artist in Portuguese art education, for example, embodied the hope of national progress that simultaneously inscribed eugenic theory about human types: the genius, idiot, insane, and normal (Martins, 2013). American music education as well was brought into the school in the hope of Americanizing Irish immigrants and newly free slaves coming into Boston schools. Children's singing and appreciation of music, for example, were to lessen concern about the threat of moral decay and degeneration of a child who was not "civilized" (Gustafson, 2005). Teaching proper songs was to remove the emotionalism of tavern and revival meetings and to regulate the moral conditions of urban life with a "higher" calling related to the nation.

The translation principles in the formation of the school subjects were then and today bonded to the educational psychological sciences. These psychologies of the 20th century and today are not concerned with understanding the production of knowledge in disciplinary fields, such as sociology, physics, and the arts (see Rose, 1989; Cohen-Cole, 2014). As with G. Stanley Hall and Edward L. Thorndike, discussed earlier, the educational psychologies embody comparative styles of reasoning to govern conduct.

It is possible to say that the translations embodied in mathematics and sciences are sublimated to the styles of reasoning generated through the psychologies of the child and the administrative principles that order teaching. This transmogrification is evident when American curriculum standards in mathematics and music education are examined (see Popkewitz & Gustafson, 2002). They do not vary significantly across school subjects. The curriculum standards of American music curriculum are fundamentally the same, deploying similar psychological distinctions about the ability to participate through informed decision-making or problem-solving, developing skills in communication (defending an argument and working effectively in a group), producing high-quality work (acquiring and using information), and making connections with a community (acting as a responsible citizen).

It is this historical question of making kinds of people and difference that is central to Diaz's *The Paradox of Making In/Equality*. The book engages thinking about how the paradox of the very categories and distinctions to secure social commitments through schooling embodies a comparative system of reasoning that differentiates, divides, and excludes in the efforts to include. It makes apparent how contemporary reforms and reform-oriented research elide the very principles governing the making of kinds of people that inscribe difference. The book poses challenges to contemporary teacher and teacher education research and reforms that seek to find more effective, high-leverage practices. Its systematic historicizing of the present opens up new pathways to think about educational research and the limits of contemporary frameworks in relation to social commitments of education.

Thomas S. Popkewitz
University of Wisconsin-Madison
March 19, 2017

Note

1. The industrialization embodied both visions progress and degeneration. Both capitalists and socialists held utopian qualities associated with the factory, for example.

References

Cohen-Cole, J. (2014). *The open mind: Cold war politics and the sciences of human nature*. Chicago, IL: The University of Chicago Press.

Cruikshank, B. (1999). *The will to empower: Democratic citizens and the other subjects*. Ithaca, NY: Cornell University Press.

Foucault, M. (1979). Governmentality. *Ideology and Consciousness, 6*, 5–22.

Gustafson, R. (2009). *Race and curriculum: Music in childhood education*. New York, NY: Palgrave Macmillan.

Popkewitz, T. (2008). *Cosmopolitanism and the age of school reform: Science, education and making society by making the child*. New York, NY: Routledge.

Popkewitz, T., Diaz, J., & Kirchgasler (Eds.). (2017). *A political sociology of educational knowledge: Studies of exclusions and difference*. New York, NY: Routledge/McMillan.

Popkewitz, T. & Gustafson, R. (2002). The alchemy of pedagogy and social inclusion/exclusion. *Philosophy of Music Education Review, 10*(2), 80–91.

Rancière, J. (2007). *Hatred of democracy* (S. Corcoran, Trans.). New York, NY: Verso.

Rose, N. (1990). *Governing the soul: The shaping of the private self*. London: Routledge.

Wagner, P. (1994). *The sociology of modernity*. New York, NY: Routledge.

Acknowledgements

From its conception, this book would not have been possible without the support of the series editors, Thomas Popkewitz and Daniel Tröhler. I am particularly grateful for the intellectual savvy of Thomas Popkewitz, who endured multiple revisions and helped me find some clarity amidst chaos. I would also like to thank Karen Adler at Routledge for her patience and guidance during this process. More thanks than I can give go to Jesse and Ada, who have tolerated my need for time.

Chapters 1, 3, and 4 are revisions of chapters or articles that are reprinted with permission and appeared as the following:

Diaz, J. (2014). Governing equality: Mathematics for all? *Journal of European Education: Issues and Studies*, 45, 3, 35–50. Reprinted with permission of Taylor & Francis LLC.

Diaz, J. (2015). Back to the basics: Inventing the mathematical self. In Popkewitz, T. S. (Eds.), *The Reason of schooling: Historicizing curriculum, pedagogy, and teacher education*. New York, NY: Routledge. pp. 184–199. Reproduced with permission of Routledge in the format Book via Copyright Clearance Center.

Diaz, J. (2017). New mathematics: A tool for living the modern life, making the mathematical citizen, and the problem of disadvantage. In Popkewitz, T., Diaz, J., & Kirchgasler, C. (Eds.), *A political sociology of educational knowledge: Studies of exclusion and difference*. New York, NY: Routledge. pp. 149–164. Reproduced with permission of Routledge in the format Book via Copyright Clearance Center.

The poem in Chapter 6, *After awhile making a proof is like making a calculation,* by Marion Deutsche Cohen from *Crossing the Equal Sign* (Plain View Press, 2007) is reproduced with permission of Plain View Press.

1 Introduction

The Study of School Mathematics Is Not About Mathematics

As an approach to studying the school curriculum and its reforms, this book takes mathematics education in the United States as its focus. But it is not only about math.[1] It is about how the curriculum embodies cultural and political dimensions that order what is known, thought, and said about who children are and who they should be. These dimensions are studied as the "reason" of schooling—or rules that historically produce notions of learning and the child in ways that are culturally and politically significant beyond school.

The book challenges the commonsense ideas of schooling and change embedded in contemporary math reforms. It weaves between discussions of the norms and standards that shape the curriculum, children, and pedagogy to bring into focus how learning in schools has little to do with content knowledge. This calls attention to what schooling does to order who children are and should be in the (re)making of curriculum.

Broadly the book considers how efforts to reform the curriculum are historically bound with projects to remake society. Specifically, it provokes how the curriculum and its reforms embody notions of social change, equality, and inclusion that present a paradox whereby efforts to reform *for all* produces notions of difference, inequality, and exclusion.

Although the focus is not merely on mathematics, but on the remaking of curriculum, math becomes an interesting site to examine the cultural and political dimensions of schooling. As a common sense of contemporary reforms both in and outside of the United States, math and science are envisioned as tools to engineer the future through the renovation of society. This promise becomes visible in statements about the importance of science and math to "the health and longevity of our Nations' citizenry, economy, and environmental resources" (Holdren, 2013). This logic that math and science sit as the king and queen of educational reforms is expressed in large-scale studies, such as the Trends in International Mathematics and Science Study (TIMSS). Hailed as the assessment of math and science knowledge of students around the world, TIMSS claims to measure and compare the achievement of students around the world. This data is deemed useful in telling a truth about what should be done through school reforms to improve achievement. Yet the test does much more than that. It expresses the belief that remaking education can reshape a nation and its future.

Present reforms in education acknowledge the social goals embodied in school learning. A recent publication from the Programme for International Student Achievement (PISA) series, *Equations and Inequalities: Making Mathematics Accessible to All* (OECD, 2016), discusses the problem of inequalities in "opportunities to learn" mathematics. It puts forth a solution for "raising disadvantaged students' opportunities to learn mathematics concepts and processes [that] may help reduce inequalities" (p. 13). Expressed as a logic about making *math for all*, which will be further discussed throughout the book, mathematics becomes a way to reason about the world, wherein opportunities to learn it presumably diminish inequalities in school and society. Embodied in this hope is a way of reasoning about how mathematics tells a truth about an inequitable world and gives children their place in it.

To highlight how the reforms of school mathematics in the United States embody political and cultural dimensions, the book unfolds how the ways of reasoning about the curriculum become visible in the practices of schooling. One such practice, explored throughout the book as a process of "alchemy," focuses on how the translation of math into math education carries psychological rules about children and teaching that seem natural. This is related to an examination of how people are made up through the curriculum as having certain qualities and traits. These traits fabricate children as either included in school mathematics or not. Relatedly, the book also provokes how the ways of reasoning about inclusion, equality, and equivalence that operate in pedagogical and curriculum models—particularly through the use of the equal sign—produce notions of exclusion and difference. It also considers how the pedagogies of teaching the equal sign in the elementary math curriculum intersect with cultural and political ways of reasoning about social equality, access, and opportunity. In pointing to the historical approach to studying the curriculum, the book invites a way of thinking about the possibilities of educational research concerned with notions of justice and inclusion beyond mathematics.

The Alchemy of School Mathematics

In school, mathematics is considered to be a universal, culturally indifferent language that models patterns and events in the world. But when math is studied in schools, it is anything but universal. This language of math and the symbols through which it is communicated carry cultural rules that organize how to learn mathematics and who can learn it.

As a different approach to research in math education, the aim of this book is to understand how the principles embodied in what is granted as a universal symbol of mathematics—the equal sign—historically produce a particular knowledge of equality, organize notions of equivalence and difference, and generate practices of inclusion and exclusion in mathematics reforms for *all* children. With *math for all* as a common sense of schooling, the book asks how it has become possible to see, think about, and act upon the knowledge of equality as a material part of the school—that gives children qualities that are comparable as either the same kinds of mathematical people or different.

Taking what is given as natural in schooling and its subjects of research and reform—*math for all*, inclusion, and the equal sign—is an approach to the study of school reform and change that provokes thinking about how curricular knowledge embodies cultural practices that order and maintain the subjects and subjectivities of schooling. This book continues in a tradition of curriculum studies that have explored the "alchemy" of school subjects (Popkewitz, 2004). The alchemy—as a process of translation that makes mathematics into math education—brings into focus the psychological models that organize the school subjects and the conditions by which they might be considered part of *math for all*. Similar studies have recently examined how math, science, and art education have historically emerged as processes of governing who the child is and should be (see Valero, 2017; Kirchgasler, 2017; Martins, 2017).

Throughout, the book will examine how the organization of school mathematics is embedded with principles of cognitive psychology as an educational science in ways that produce the image of the child as a math learner with certain traits. This ordering becomes visible through the tools and language of psychology as it assembles mathematics with social and psychological concepts, like motivation, creativity, independence, and metacognition in ways that have little to do with mathematics.

Exploring the alchemy of mathematics education is to consider how political theories of participation and equality are embedded in the principles that organize teaching through conceptions of the child. The content of the curriculum entails rules that provide certain ways of thinking about children to organize school for them. But it is not just about that. Children are not just there, as autonomous objects, waiting to be labeled, evaluated, and sorted. The alchemy embodies ways of reasoning about children that make them into certain kinds of people.

Children are not mathematicians after all. What they are to know and do assembles and connects with a reason of schooling that shapes how to think, see, and act in school as certain kinds of people—problem solvers, creative thinkers, and rational decision makers in society.

This reason about who children are and should be as "equal" participants in schooling and society is studied by making explicit the historical rules that organize what is known, talked about, and acted on in the present. Studying the alchemy of school subjects makes it possible to see not the universality of math but its historical, political, and cultural dimensions that highlight how math education has little to do with the disciplinary knowledge of mathematics.

Schooling and Making Up Kinds of People

The alchemy of school math is connected throughout the book to cultural norms that have historically defined notions of access and inclusion within reforms aimed at *math for all*. Although these norms provide a way to see and describe children and the perceived problems in school math, they are not natural ways of thinking or talking about children. Rather, these classifications fabricate, or "make up people" as certain kinds (Hacking, 2002).

As a fabrication, identities are given to children—not merely as descriptors of who they naturally are but as a way to understand and categorize what is taken as the problem of difference. This is evident in how children are given traits that are ordered on a continuum of value—from high achieving, successful, mathematical kinds of people to struggling, low-achieving, not math people. Children are not born with these fixed identities and differences. They are assigned through school and make possible certain ways of seeing, thinking about, and acting upon children.

In studies of schooling, fabrication is a sociological notion and draws attention to how schools participate in making kinds of people in two ways (Popkewitz, 2013). One, the classifications of children are *made-up fictions* that provide rules for how to think about and see them in school. Two, the classifications that are embedded in the rules about schooling also *organize who people are* and should be. These categorizations of children produce theories about who they are, how they should act, and who they should become through school.

As a way of thinking about the problems of research and schooling, the focus on fabrication makes problematic the classifications and categories that seem to provide social facts about children in schools. These facts are expressed as psychological traits and are embodied in the curriculum to manage how to think about them to act in ways that teach them how to live as certain kinds of people. In this way, the fabrication of children serves as a normalizing pedagogy.

A math learner has become the kind of person children are supposed to be to solve problems, reason, persevere, think flexibly and efficiently, communicate, and represent the mathematics they are doing. Yet, this is not simply about mathematics. These practices provide a way to categorize and classify children. The work of school mathematics relies upon and produces this image of the child as a math learner, as a historical kind of person who appears to think and act in particular ways. This kind of person has emerged to characterize what it means to learn math and frames the efforts of making all children into particular kinds through mathematics.

The interaction of educational psychology with math curriculum appears as a seemingly objective strategy for organizing knowledge of children and their thoughts. But pedagogies organized through the educational sciences of cognitive and social psychologies embody a way of comparing to *make people equivalent*. This equivalence is established by what the child should see, think, and do as a math kind of learner to solve problems involving the equal sign. Ironically, the ordering of the child's thinking produces divisions in terms of who is not that kind child, discussed throughout the book as a "double gesture." The presumed differences among children are then reduced to gaps,[2] or inequalities, and used as distinctions that rationalize further categorizations to establish some sort of equivalence. These divisions have implications for inclusion or exclusion that go beyond school mathematics and are studied here as part of the cultural politics of the curriculum.

The Double Gestures of Inclusion and Exclusion

Importantly, the book draws attention to how the fabrication of children through the alchemy of school math inscribes cultural norms that differentiate who is or is not part of *math for all*. This is political in that the psychological normalization of the child as a math learner constructs identities and equivalences among children while embodying a logic of comparison. The comparative mode of reasoning that organizes who children are and should be also entails a notion of difference.

This phenomenon is examined throughout the book in how the production of identities and differences embodies double gestures and simultaneously organizes terms of inclusion and exclusion (Popkewitz, 2008). This can be seen in how the reforms aimed at mathematics for all in the United Sates carry both the hope that a particular kind of child will contribute to the future progression of society as well as the fear of the child who threatens that progress. Whereas math in schools is and has been connected to conditions for equality and inclusion, it also entails exclusions and the production of inequalities.

The double gesture will be visible in how the logic of equality operates in the curriculum through the use of the equal sign and paradoxically circulates a way of reasoning about difference as the problem of inequality. In looking closely at attempts to make *math for all* across three periods of math reform in the United States, the analysis attends to what have become the seemingly natural ways of seeing and talking about children through traits that inscribe similarities and differences.

This paradox of in/exclusion is not intentional in reforms or part of what has been called a hidden curriculum. Rather, it is an effect of power that becomes visible in the historical ways of reasoning that produce identities and differences that both include and exclude. Throughout the book, these identities and differences are not taken for granted as they have become the very limits and conditions under which efforts to organize children's math learning emerged, are sustained, and might change.

By calling into question how the consistent application of psychological rules to the improvement of education often reinscribe difference as the problem the reforms seek to eradicate, this study examines how educational problems are conceived, how reforms and their purposes are given intelligibility, and how objects of research are historically contingent and defined. In this way, the study ties the alchemy of school mathematics, the making of children as certain kinds of people, and the double gesture of inclusion and exclusion to a broad examination of how the historical conditions that produce ways of reasoning about all children enclose and intern possibilities for curriculum reforms aimed at notions of social justice, equality, and inclusion.

The Equal Sign as an Actor

Given this central focus on the relation between the math curriculum and social theories of equality and inclusion, the book argues that the use of the equal sign in elementary school mathematics is not about the alleged cultural indifference of mathematical thinking but relates to historical, political, and social understandings of equality, particularly in reforms aimed at *math for all*.

As the equal sign appears in the school math curriculum and connects to calls for *math for all* in the United States from the 1950s to the present, it orders children's thinking about equality in mathematics by assigning identities to things of the world—as expressions of equivalence and nonequivalence. In this way, its use in the school also embodies cultural ways of reasoning about equality, equivalence, and difference that go beyond dimensions of seemingly pure mathematics. Both the sign and the comparative logic it embodies organize the curriculum and the broader understandings of in/equality in math education. This ordering of thought about equality is central to reasoning about how children are to participate in school and American society as "equals."

Placing the equal sign and its relation to notions of equality at the center of analysis reveals how threads of thought weave together to historically produce and maintain identities of children as certain kinds of people. As such, the equal sign is treated as an historical actor, framing and framed by particular ways of knowing and being in the world. How the equal sign is used to construct knowledge, organize pedagogical practices, and order children's identities is related to notions of equality and frames questions of what constitutes inclusion and justice in math education for all.

The analysis of the relationship between the equal sign as an actor and notions of social equality is not to foreclose the possibilities for human action and agency. However, autonomous actors are not positioned as the cause of change or the makers of history. Historical changes are attributed to overlapping ways of knowing about and organizing equality that shape the historical ways of reasoning about in/equality.

This history will not be told as an account of how people have acted to change the shape of school mathematics. Instead, historical actors are thought of as "conceptual personae" (Deleuze & Guattari, 1991). Bringing specific authors into the narrative attends to how they stand as figures of thought of particular modes of reasoning about children and their learning.

In Chapter 4, for example, the works of Jean Piaget and Jerome Bruner are discussed in the analysis of how the child's reasoning about equivalence was related to notions of social progress and citizenship planning during the period of math reform that followed World War II. The reading of their texts is not to understand their intentions but to see how their work was imported into the American curriculum to articulate a new kind of child. Considering the authors as conceptual personae, the analysis uncovers how principles of cognitive psychology assembled with the math curriculum to inscribe cultural norms about children's mathematics and their learning.

Decentering the autonomous subject and viewing the equal sign as an actor assume a particular notion of change. The actions to be made and the changes that count as reform are considered within a larger system of reasoning, and "it is through locating changes in the rules of reason that we can think about change" (Popkewitz, 2001, p. 168). That is, in exploring the rules that organize equality as a cultural practice, new possibilities for seeing, knowing, and acting may emerge that do not place the problem of in/equality in the child.

A History of Reform: Chapter Outline

Framed by the concern that equality is taken as neutral while embodying ways of reasoning in the curriculum that unreasonably mobilize action, the book works to unsettle the logic of equality. It will examine how notions of equivalence and difference are embedded in *math for all* reforms and tied to learning the equal sign as a fact of the curriculum. This reading of the curriculum offers a way to see how the logic of in/equality inscribed in the equal sign is given meaning and conditioned by particular social, cultural, and political phenomena. This becomes evident in how the ways of teaching and learning equality through the equal sign 1) give meaning to the inclusive premise that *all children can and should learn mathematics* and 2) are modified by historical conditions of *if, when,* and *because* that organize in/exclusion in *math for all.*

With the present as a starting point, Chapter 2 begins by discussing a particular fact of schooling that is taken for granted—the teaching and learning of the equal sign. This introduces contemporary math education reforms aimed at *math for all* as a site of the cultural politics of school subjects. Continuing to give attention to alchemy, fabrication, and double gesture, Chapter 1 names the problem of educational reform and research as one of examining the paradox of in/equality in the movements toward *math for all* as a way of reasoning about the politics of representation, equivalence, and difference.

Bringing a different approach to math education and reform research that focuses on the reason of inclusion and diversity, Chapter 2 opens the possibility to move research in math education beyond its current structural and institutional limits. Running against the grain of studies aimed at producing equity or recognizing diversity, the book works to interrogate, rather than assume, questions about equity and equality in mathematics. Not taking as granted the role of school math in providing all individuals with an opportunity to become mathematical kinds of people, Chapter 2 continues to explore the relationship between the knowledge of schooling, making up children, and the paradoxical logic of organizing inclusion and equality in schools.

Unraveling the commonsense ways of reasoning that are used to think about children and their math learning, the book moves back and forth from the present and the past. The approach to understanding the rules that organize contemporary *math for all* reforms is articulated in Chapter 3 as a "cultural history of the present" (Popkewitz, Franklin, & Pereyra, 2001). This study of school math gives attention to how the norms about equality, equivalence, and difference—as frameworks for thinking about inclusion—are historically constituted and brought into the present. It follows a tradition of curriculum studies that looks at how the historical subjects of schools are fabricated and come into the present as cultural norms that shape schooling and its reforms (see also Lesko, 2012; Ramos do Ó, Martins, & Paz, 2013). The historical work seeks to understand how the logic of equality, equivalence, and difference is embedded in the curriculum in ways that classify and divide children.

This approach to studying math educational reform does not provide a linear and causal narrative as the effect of historical agents and intentions. Rather, it looks broadly at the historical ways of reasoning about children and their math learning that operate as the logic of reforms. Chapter 3 attends to how children's math education has become intertwined with concerns for social inclusion and progress. This becomes evident in the discussion of the historical conditions that have made *math for all* reasonable as a way of organizing elementary math reforms. These conditions importantly emphasize a post-World War II shift in the rationality given to mathematics, the social sciences, and cognitive psychology that assembled to situate children and elementary math education as integral to the planning of society.

Taking the postwar turning point as a historical marker, Chapter 4 examines the reorganization of a new math curriculum that emphasized children's learning of equality and equivalence as foundational to learning mathematics. This placement of the structures of equivalence and equality was taken as central to all students using mathematics as a tool for living a modern life. Beyond mere mathematics, learning equivalence and equality was to be understood within a broader cultural frame whereby the curriculum was (re)organized according to principles of national progress in postwar America, notions of social planning, and aims for educational equality.

Drawing upon cultural history, political theory, policy documents, and an educational psychology that presumed children could master the content of mathematics given its presumably universal structures, Chapter 4 examines how a new curriculum movement produced new ways of thinking about the relationship between the child's development in mathematics and the progress of the nation as an effect of a mathematical form of creativity, flexibility, and independence—given as traits of an intelligent citizen. This identity given to children's mathematics and norms of participation as citizens established a relationship that defined the terms by which to see the child who appeared as disadvantaged and stood as the reason for inequality.

With children's mathematics tied to notions of citizenship and social progress, it became possible to further distinguish who would or would not be part of *math for all*. Chapter 5 examines how mathematics learning was studied as a way to achieve excellence in math education for all through the examination of what were taken as individual motivations, abilities, and interests in learning math during the 1970s back-to-basics movement. The inscription of individual differences in the mathematics curriculum is emphasized in how learning basic operations was a skill—not only for making sense of and doing mathematics but also for living as an individual distinct from others.

Analyzing school math problems translated through a psychology that assumed children as naturally capable of being equally, but differentially, motivated, Chapter 5 examines how the use of the equal sign in the curriculum produced cultural narratives of the child as a self-responsible, -reliant, -directed and -motivated individual, as distinguished from what emerged as a mathematically unmotivated and disabled child.

The shifting historical significations that mark a child's inclusion into *math for all* continue to be explored in Chapter 6. As back-to-basics reform shifted into a standards-based movement during the early 1980s, the image of the child as a mathematical kind of person emerged. The alchemy is made visible through the use of the equal sign in organizing math power, problem-solving, and mathematical literacy as equivalent to the normalized social and psychological traits that every person was presumed to have. Research in mathematics education and changing psychologies that emphasized the awareness and communication of one's thinking assembled to give characteristics to the child's development of this power while also producing mathematical illiteracy as a problem of difference and inequality.

With the norms of identity that are to include all children into the *math for all* shifting from ideals of citizenship, to distinctions of individuality, to standards of personhood, Chapter 7 discusses the historically shifting identities and differences that continuously redefine the terms of inclusion and exclusion. Attending to the double gesture, discussed previously, it highlights how the images of the child as a mathematical kind of citizen, self, and person produce the distinctions that represent inequality as a problem of difference.

Situated in current reforms that take school subjects as the origin of change, Chapter 7 interrogates the contemporary emphasis on teachers as content experts and research that aims to prescribe best practices. This is to question the limits of these reforms—as resting upon seemingly fixed ways of knowing and seeing children. Operating within a way of reasoning about inclusion and inequality that is historically contingent, the identities and differences of the children who can and cannot be part of *math for all* will continuously shift.

A moving target, in perpetual motion, the relationship between the content and pedagogy is discussed as a way to study the curriculum and provide an opening for change that does not place the problem in the child. Chapter 7 concludes by invoking a broader discussion of how this research contributes to the study of schooling and curriculum, particularly in relation to issues of difference, diversity, inclusion, and equality. Emphasis is given to a consideration of how studying the curriculum as a cultural history of the present can provide alternative ways of considering the limits of reform and open possibilities for change.

Looking at the politics of inclusion and exclusion at the site of mathematics education reform in the United States offers possibilities to explore the cultural and historical production of the boundaries that have seemed to naturally characterize the child as a certain kind of person.

However, it does not presume to provide the framework from which all other math education reforms unfold—nationally or internationally. The United States is not a case that can be generalized and treated as a universal. Nor is it the singular example of the intersection between math education and the politics of inclusion and exclusion. Although the analysis focuses on how the (re)making of the curriculum comes into play with social, political, and cultural questions about making children into certain kinds of people, it does not take the United States as the only context in which this occurs. Whereas math education has cultural dimensions beyond the United States, here it is treated as one possible site to explore the cultural politics of making schooling for all.

The book offers a way of rereading the math curriculum in relation to cultural, historical, and political dimensions outside of mathematics—but the broad scope moves the discussion beyond school math. In considering the politics of the representation of identity and difference and the production of in/equality, the book provokes a way of reading what the curriculum does that moves beyond the surface of discussions about equity, diversity, and inclusion. For educators and researchers, it begs questions about how the subjects of school reform are historically produced; how they might be seen, studied, and thought about as an engagement with the politics of schooling; and how the possibilities for change are hinged upon understanding the common sense of the present as open to question.

Notes

1. Throughout the book, the colloquialism "math" is used as an abbreviation of mathematics, recognizing that "maths" is also used in various parts of the world.
2. The "achievement gap" is often used to characterize the discrepancy and inequality in student math performance. With divisions drawn along gendered, linguistic, economic, and racial lines, the "gap" is described and explained as an effect of the differences between the students. So as to not locate the problem

of achievement in the students, this study examines the historical, cultural, and political factors that shape how achievement is determined as an effect of power.

References

Deleuze, G. & Guattari, F. (1991). Conceptual personae: Introduction: The question then . . . (H. Tomlinson and G. Burchell, Trans.). In Deleuze, G. (Ed.), *What is philosophy?* New York, NY: Colombia University Press. pp. 61–84.

Hacking, I. (2002). Inaugural lecture: Chair of philosophy and history of scientific concepts at the College de France. *Economy and Society*, *31*(1), 1–14.

Holdren, J. (2013). Federal Science, Technology, Engineering, and Mathematics (STEM) Education Strategic Plan. A Report from the Committee on STEM Education National Science and Technology Council.

Kirchgasler, K. (2017). Scientific Americans: Historicizing the making of difference in early 20th-century U.S. science education. In Popkewitz, T., Diaz, J., & Kirchgasler, C. (Eds.), *A political sociology of educational knowledge: Studies of exclusion and difference*. New York, NY: Routledge. pp. 87–102.

Lesko, N. (2012). *Act your age! A cultural construction of adolescence*. 2nd Edition. New York, NY: Taylor & Francis.

Martins, C. (2017). From scribbles to details: The invention of stages of development in drawing and the government of the child. In Popkewitz, T., Diaz, J., & Kirchgasler, C. (Eds.), *A political sociology of educational knowledge: Studies of exclusion and difference*. New York, NY: Routledge. pp. 103–116.

Organization of Economic Cooperation and Development (OECD). (2016). *Equations and inequalities: Making mathematics accessible to all*. Paris: PISA, OECD Publishing.

Popkewitz, T.S. (2004). The alchemy of the mathematics curriculum: Inscriptions and the fabrication of the child. *American Educational Journal*, *41*(4), 3–34.

Popkewitz, T.S. (2008). *Cosmopolitanism and the age of school reform: Science, education and making society by making the child*. New York, NY: Routledge.

Popkewitz, T.S. (July 2013). The sociology of education as the history of the present: Fabrication, difference and abjection. *Discourse: Studies in the Cultural Politics of Education*, *34*(3), 439–456.

Popkewitz, T.S., Franklin, B., & Pereyra, M. (2001). *Cultural history and education: Critical essays on knowledge and schooling*. New York, NY: RoutledgeFalmer.

Ramos do Ó, J., Martins, C., & Paz, A. (2013). Genealogy as history: From pupil to artist as the dynamics of genius, status and inventiveness in art education in Portugal. In Popkewitz, T.S. (Ed.), *Rethinking the history of education: Transnational perspectives on its questions, methods, and knowledge*. New York, NY: Palgrave. pp. 157–179.

Valero, P. (2017). Mathematics for all, economic growth, and the making of the citizen-worker. In Popkewitz, T., Diaz, J., & Kirchgasler, C. (Eds.), *A political sociology of educational knowledge: Studies of exclusion and difference*. New York, NY: Routledge. pp. 117–132.

2 The Cultural Politics of School Math

The Problem of In/Equality in the Equal Sign

The equal sign figures importantly in the ordering of elementary school mathematics in the United States. Its placement as a focus of math reform and research makes that clear. Take, for example, the importance given to children learning equality as a relationship of sameness, represented by the equal sign (Carpenter, Franke, & Levi, 2003; Falkner, Levi, & Carpenter, 1999). Learning the equal sign seems integral to children's understanding of mathematics.

But this learning is about more than mathematics. The equal sign is embedded in a *math for all* movement in which math learning is to be accessible to and inclusive of every student (Choike, 1996; Frankenstein, 1995; Oakes, 1985; Secada, 1995). Learning mathematics is to produce social equality—expressed as *all* students having the same opportunity, access, and possibilities in school and in life as *equals*. This can be seen in *An Agenda for Action*, where the National Council for Teachers of Mathematics (NCTM, 1980) calls for a higher level of mathematical skill and competence for *all*. In another influential document, the *Curriculum and Evaluation Standards for School Mathematics* (NCTM, 1989), "opportunity for all" was formulated as one of four societal goals for mathematics education. These reform projects assume that social, economic, and educational inequalities can be minimized if *all* children have the opportunity to learn mathematics.

The logic underpinning of this research and related reforms is that with a foundational knowledge of equality and equivalence in mathematics—expressed in the equal sign—children will have a better understanding of equality as a relationship, greater access to higher levels of mathematics, more academic opportunities, and an overall improvement in economic and social standing. Circulating within this way of reasoning, the equal sign and the child's understanding of equality are represented as neutral tools to achieve that aim.

Looking at the use of the equal sign in the elementary curriculum, the chapters throughout the book explore how the practices for learning equality carry cultural rules of equivalence and difference—as the organizing principles for comparing children and things. In this reading of the curriculum, the concern is with how the logic of equality makes it possible to classify and compare children as certain kinds of people. The rules for who get to count as part of *math for all* are coupled with the impulse to provide opportunities and access to mathematics. Yet the cultural coding of equivalence that shapes who gets access to math as the embodiment of justice and equity are often taken as neutral, apolitical, and ahistorical. This limits the very efforts of the reforms by claiming differences in children as the explanation for inequality. Most problematically, it places the problems of school mathematics and its inequalities in children.

Considering how the logic of equality embodied in the use of the equal sign in the school curriculum order notions of identity to manage inequality as differences among children, this chapter introduces the paradox of in/equality. Situated in efforts to make *math for all* in the United States, the study of this paradox examines how the boundaries of inclusion and exclusion are established in the (re)organization of the elementary math curriculum. This chapter argues that the rules normalizing how children should think and act as math learners are embodied in the curriculum by examining how the principles of educational psychology function to establish equivalences as modes of comparing all children. It attends to how the pedagogies organized through the educational sciences of cognitive psychology produce a way of comparing children to *make equivalence and establish differences.*

If math is to be (re)made for all children, then an attempt to resituate the problem of in/equality, in/exclusion, and in/justice in mathematics education is necessary. The examination here

> is not a matter of taking pleasure in paradox, but of fundamental inquiry into the meanings of knowledge, teaching and learning: not an amusing journey into the history of pedagogy, but a matter of timely philosophical reflection on the way in which pedagogical logic and social logic bear on each other
>
> (Rancière in Bingham & Biesta, 2011, p. 1)

in the making of school mathematics for all. Looking at the relationship between the pedagogical and social logic of equality, this study is both about math education and it is not. Framed by the concern that equality is granted as neutral and logical while embodying ways of knowing in the curriculum that produce difference as inequality, the aims go beyond school mathematics to more broadly examine how the historical conditions that define *all children* impose limits on curriculum reform efforts aimed at notions of social justice and inclusion.

Reforming the Curriculum for All and Rules of Reasoning About Equality

During the last two decades of the 20th century, clear reference was made to notions of equality, equity, and inclusion in math educational reforms in the United States. This *math for all* movement has placed an emphasis on raising the expectations for all students to learn and excel in mathematics.

As part of the *math for all* reforms in school math, focus is placed on granting equal access and opportunity to all students—"regardless of personal characteristics, backgrounds, or physical challenges"[1] (National Council of Teachers of Mathematics (NCTM), 2000, p. 12). The movement embodies notions of equal opportunity and outcomes related to a notion of equity. Equity, embedded in concerns about social equality, generally includes three aspects: equal opportunity to learn mathematics, equitable treatment based on the perception and description of being equal, and equal educational outcomes (Fennema & Meyer, 1989). Equality and equity seem to be implied in one another and have both been taken as granted in the discussion and redirection of school mathematics in the present.

Here it is not important to distinguish between notions of equity and equality. What is at issue is how a "logic of representation" (Olkowski, 1999) orders the *ways of thinking about and acting upon equality and equity that assume the identity of an individual, represented as distinct from and comparable to others.* This comparison is made possible given a logic wherein similarities can be established among things and people through which differences are categorized. As related to the commitments to equality and equity already mentioned, both seem to presume the representation of people and their worlds as fixed and, therefore, possible to identify and compare. Considering how the logic of representation and equivalence operate in the ordering of school subjects to produce cultural norms of identity and difference is a central focus of the book.

The cultural norms of sameness and difference that determine notions of access, achievement, and inclusion within *math for all* provide a comparative reference by which to see children in relation to others. This is not simply a way of thinking or talking about children. These categories fabricate, or "make up people," as certain kinds (Hacking, 2002). As a fabrication, identities are given to people—not merely as descriptors of who they naturally and already are but as a way to understand and categorize what is taken as the problem of difference.

When this logic of equality functions in the school to organize notions of equivalence and difference, children are made to have certain traits—shaping how they are seen, thought about, and acted upon and how they see themselves. In this way, looking at how the use of the equal sign carries cultural inscriptions of equivalence and difference is a move to explore the formation of the subjects in the math curriculum. The identities that define who children are and who they should be as math learners are not natural and given. Instead, the identities are contingent as effects of shifting cultural conditions and historical factors—always open to question.

Opening up the characteristics that define children's mathematical thinking—creative/close-minded, rational/nonsensical, flexible/rigid, and abstract/concrete—makes visible how the child is fabricated through the reform of school mathematics. This raises "questions about the constructed nature of experiences, about how subjects are constituted as different in the first place" (Scott, 1992, p. 25) and disrupts the very distinctions through which experiences of inclusion and exclusion in school are seen, thought about, and acted on.

Interrogating the politics of identity, norms of representation, and rules for in/exclusion is to understand how identities and difference are constructed and inscribed on children as characteristics that seem useful for defining children as mathematical kinds of learners. Given as a kind of child the curriculum assumes, the math learner is historically defined by a series of rules, standards, and practices. One such practice—learning the equal sign—has become central to organizing ways of reasoning about of equivalence and difference as the expression of in/equality.

This is not only a mathematical idea but is embodied in the subjects of the school as cultural practices that govern how to see, think, and act. In this way, the equal sign orders and maintains a system of reasoning that organizes cultural norms of equality as rules by which people are to live (Foucault, 1991). This reasoning assembles with and resembles the cultural principles that organize notions of equivalence and difference by which to determine who is equal and who is not, marking the boundaries of in/exclusion.

Exploring the relation between the equal sign and the organization of equality in the contemporary *math for all* reforms in the United States treats the equal sign as a cultural practice. More than a symbol of mathematical logic, the rules for learning equality intersect with certain truths about children and organizes how they should *all* see, think, and act in the world. Said another way, "mathematics is a discipline that disciplines" (Phillips, 2015, p. 6). The next sections will further examine how the rules of equivalence and difference that order children and their learning are embodied in the equal sign.

The Equal Sign as a Modern Fact

Contemporary math reforms in the United States expect that all children should "understand the meaning of the equal sign, and determine if equations involving addition and subtraction are true or false" (NGA & CCSSO, 2012). The logic of the curriculum organizing knowledge of equality is related to how the equal sign functions as standard content for learning. The equal sign itself can be considered a "modern fact" (Poovey, 1998) of the curriculum in how it organizes knowledge of equality that every child is supposed to learn.

As a fact of the elementary school curriculum, the equal sign is rarely seen as anything other than a symbol of mathematical thinking. Yet it acts within historical and cultural ways of reasoning about mathematics, schooling, and children that intersect in the social spaces of school. The sign is to be learned in relation to the psychologies of learning that translate mathematics into something that children can use to think about and model equivalence and nonequivalence. The process that translates mathematics into math education is important as it generates cultural ways of thinking and acting that are never merely about mathematics. This translation can be thought of as "an alchemy"—or a transformation of the content of mathematics through the language and tools of educational psychologies (Popkewitz, 2004). Discussed in the introduction, such translations have been taken as natural and necessary because children are not mathematicians. But in both the use of math as a way of modeling the world and educational psychology as a tool for mapping the mind are particular principles about how to quantify, measure, and compare children as certain kinds of people.

As a modern fact, the equal sign has embodied a social and cultural way of reasoning about equality. Organizing knowledge about how to see the world, and the child's place in it, the equal sign and its translation into the curriculum embody cultural ways of reasoning about equality. So, as the equal sign comes into the elementary school curriculum, mathematical notions of equality interact with the taken-for-granted psychological rules to shape the practices and norms of how children are to learn in ways that order who they are and who they should be.

The use of the equal sign in school math interacts with the rules of reasoning about equivalence and equality that operate in the curriculum to order two notions: 1) the rationality of mathematics as a field of thought that models the world and 2) the logic and learning of mathematics through educational psychologies as the presumed sciences for knowing the child.

A closer look at how the use of numbers, symbols, and objects come together with knowledge about children's thinking bring to light how the logic of equality embodied in the equal sign are about so much more than mathematics.

Mathematical Objects and Styles of Reasoning

In school mathematics the equal sign rarely appears alone. It makes sense in relation to other symbols and objects. The equal sign and numbers organize problems for children to solve about identity and difference from which to understand the fact of equality. When children are asked to determine this equality, they are to see and think about the numerical statements of equivalence as a representation of some truth in the world.

Are these math sentences true (T) or false (F)? Why?

$$4 + 5 = 9$$
$$7 = 3 + 4$$
$$8 = 8$$
$$10-5 = 2 + 3$$
$$8 + 2 = 10 + 4$$

(Beatty & Moss, 2007, p. 31)

The numbers and the equal sign appear to establish equivalences and difference, giving meaning to the expressions as either true or false. Here the numbers do not make sense in and of themselves. They are validated as social facts and involved in constructing the social world, not just representing it (Porter, 1995). In school math problems, like those here, the numbers work as a strategy for constructing a truth about equivalence that is seen as a mathematical form of reasoning. This highlights how the use of "numbers in the representation of the world is predicated upon procedures of classification and separation of the 'identical' and 'different' which results in the building of a rigid perception of reality" (Patriarca, 1996, p. 9). More than mathematics, the equation also embodies rules for how to see difference, sameness, equalities, and inequalities as a form of knowing.

The practice of classifying statements by their truth value embodies a way of categorizing relationships in the world. The problem-solving here can be considered to inscribe in the child a "style of scientific reasoning" (Hacking, 2002). Carrying a way of thinking, the expressions of in/equality carry rules that define equality as a relationship of sameness. These rules then are to be internalized and applied in a presumably rational way to other instantiations of equality represented with the equal sign.

The use of numbers and symbols that are given validity in modeling the world, its patterns, and relationships also create that world and organize how it is to be seen. Through mathematics, children are to internalize behaviors whereby the world can be calculated, patterns can be followed, and order can be maintained. In this association, children's sense of numbers—what they represent, how they behave and relate, and how they can be manipulated—organizes the sense of order and calculability that is internalized through school math. Numbers and the logic given to them in modeling and organizing things and events carries with it certain expectations for how one should live in relation to those things.

This sense provides the rules for how to establish equivalences by comparing certain things by giving them an order, determining patterns, and assigning classifications. So children both learn about and become objects of a style of reasoning that is expected to make them more rational than they presumably are. This way of thinking about mathematics situates school math as a tool to move children toward reasonable thought as a way of living and planning a better life. Similar discussions frame *The Mastery of Reason: Cognitive Development and the Production of Rationality* (Walkerdine, 1988), wherein the focus is on how notions of rationality and reason are thought to be produced in children through learning school mathematics. This relationship between children's learning and math as a form of reasoning for planning the future is explored further in the next chapter.

Where the *what* of school math intersects with *how* children are thought to learn it is the human who the curriculum assumes is defined—as the math learner, the (number) sense maker, the rational thinker, the math person. These modes of thought and participation—particularly in relation to learning about numbers—are attributed to particular abilities, intuitions, awareness, feelings, and innate capacities (Berch, 2005). Here a child's learning about what sense to make of numbers is explained in the language of learning sciences in ways that seem natural and neutral. The content and pedagogy of the curriculum assemble to determine what kinds of people get to count as math learners by assigning traits to children. Children are not born to see and think (or to be seen and thought about) in this way. The problems of mathematics order certain rules and fabricate the child's modes of thinking and participating in school through an interaction with and translation through psychological knowledge about children.

Establishing Equivalence: Developing the "Reason" of Equality Through the Learning Sciences[2]

Together with the language and logic of mathematics, the learning sciences organize and classify how children can embody the rules of thought instantiated in the equal sign: how to think about equivalence, how to see difference, and how to make comparisons through an evaluation of sameness. In school math, this logic of equality intersects with learning theories that organize how children should learn these rules. The learning sciences, then, have become central in assigning characteristics and traits to define who children are and who they should become through school math.

Throughout the book, the discussion brings to the fore how the psychological rules intersect with expectations about who children are and how they should learn. For example, the psychological principles that organized what it meant to be a problem solver in the standards-based reform efforts (Chapter 6) were embodied in the school math problems as rules for social behaviors and expectations of self-awareness and confidence. In the new math reforms (Chapter 4), the psychological principles of creativity, independence, and abstraction gave meaning to what it meant to learn equality as well as provided the rules for participating as citizens. This is the "alchemy" of mathematics, translated into the curriculum through the psychological norms that order how children are to internalize the rules of reasoning about mathematics, the world, and their particular place in it.

The translation of math into mental, social, and behavioral expectations for children does not just happen. It has become possible through the historical emergence of psychology as the social technology for translating disciplinary knowledge into the schools. Psychology has not always been considered useful in providing a lens through which children and their learning would be visible. This can be seen in historical discussions of how attempts to professionalize, institutionalize, and legitimize the field of early childhood education in the first third of the 20th century in the United States were tied to assumptions about making it more "scientific" through educational psychology (Bloch, 1987). Through the work of psychologist G. Stanley Hall, child study grew in prominence and provided a way to observe, record, quantify, and predict children's behaviors as the foundation of a "scientifically" sound curriculum design and "legitimate" education (ibid, p. 46). Psychological theories about and research on children emerged historically as a way to provide a rationalization for the work of schooling.

To talk about psychology this way is to highlight how it has become the mechanism for interpreting mathematics and bringing it into the school via knowledge about the content of children's minds and their development, processes, and abilities, which are not at all neutral. Yet when the abstraction of the mind and how it presumably works is tethered to behaviors, mental processes, and beliefs that get called rationality, then it becomes taken for granted and makes it possible to "see" learning in the content areas. Although the psychological norms about how children think, act, and learn in the world are taken as the dominating translating tool for curricular and instructional reforms, the rules and principles that order the sciences of educational psychology as well as its objects of research are not natural. But because children are not mathematicians, this knowledge of who they are and how they learn is taken as integral to making them into mathematical kinds of people!

Mathematics becomes math education through its translation into the curriculum through the rules of learning sciences. This transformation of the disciplinary knowledge into the school content carries cultural rules and norms that are to be internalized by children and demonstrated as learning. In this translation, the style of reasoning about equality and equivalence embedded in the equal sign provides a way to govern how a child should think and act—as a rational, reasonable, and mathematical kind of learner. This rationality ascribed to mathematics, as a tool of thought, organizes what comes to count as rational. This is clear in how it has become possible to classify children and their thinking through the language of the learning sciences. The "mathematically disabled" child would not be possible to see or think about without the measures of intelligence, personality, and ability that give this classification meaning through the language and tools of the learning sciences.

The classification of rationality and the style of reasoning it presumes interact with notions of equivalence to organize a seemingly scientific knowledge of learning. Without this logic of equivalence, comparisons cannot be made among children as factors of sameness and difference. Within this comparative cultural framework, the child can judge and be judged in terms of experiences and competencies in relation to others. An article titled "Teaching the Meaning of the Equal Sign to Children With Disabilities: Moving From Concrete to Abstractions," provides a way of seeing how this comparative rationality is inscribed in the use of the equal sign (Beatty & Moss, 2007). The research is to substitute children's "intuitive proto-quantitative understanding of equality" (p. 30) to "support the development of a more sophisticated and mathematically useful understanding of the equal sign" (p. 28). Translating children's understanding into an objective knowledge with numerical symbols and objects, the project works to represent what children apparently intuit but misunderstand. The aim, then, is to "find efficient technologies that replace children's 'intuitive' reasoning in which misconceptions are contained with new sets of rules for acting and seeing" (Popkewitz, 2008, p. 140). Embedded in the notion of misconception is a certain truth value of equality in the world, arrived at through the learning of mathematics.

The new rules for acting and seeing require children to organize numbers, symbols, and other objects to construct a relationship of sameness. To establish equality, the child must learn that concrete objects stand in to represent abstractions of numbers. To be seen as progressing or developing, the children are to demonstrate a particular style of thinking through which they translate cultural objects (counters, numbers, and symbols) to mediate equality's meaning. What children should see and how they should see it are techniques that operate through the curriculum to order a style of thinking that the child should adopt and practice as problem-solving.

To answer "What number goes in the box?" in the problems $6 - 2 = \square$ and $5 + 7 = \square + 8$, the child is expected to see the mathematical statements as a strategy for establishing equivalence (Beatty & Moss, 2007, p. 32). That is, the child should compare what lies on one side of the equal sign to another. This problem-solving embodies both a mode of learning and a style of reasoning about how to represent and compare things as equivalent.

In this way, the expression $6 - 2 = \square$ is much more than a mathematical question. *It also organizes thought and action as a logic of equivalence to make people objects of comparison.* Within a comparative style of reasoning, it becomes possible to see children as objects of the sciences of learning psychologies on a continuum of rationality, understanding, or ability. These categories presumably identify how children should think and act to solve the problems of school mathematics. At the same time, these characteristics are taken to represent the inner workings of the child's mind and are made visible as the materialization of learning. This visibility of learning is presumably expressed as children manipulate and quantify objects, order a linear thought process, and articulate a rationale by responding (correctly) to the question: "How do you know?"

The rationality of educational psychologies is to know the child's mind and how thought processes function. In rendering the subjects of the school as intelligible and manageable, the principles of psychology work to organize the norms of participation as practices the child should internalize through learning the content of the curriculum. This interaction of educational psychology with the math curriculum appears as a seemingly objective strategy for organizing knowledge of children and their thoughts. But pedagogies organized through the educational science of cognitive psychology embody a way of comparing to *make people equivalent.*

Equivalence, as both a comparative mode of thought, is established by ordering what the child should see, think, and do as a mathematical kind of learner to solve problems involving the equal sign. Ironically, the ordering of the child's thinking also produces divisions in terms of who is not that child, discussed in the introduction as a "double gesture." The presumed differences between children are then reduced to gaps, or inequalities, and used as distinctions that rationalize further categorizations to establish some sort of equivalence.

In/Equality: Calculations of Difference and the Paradox of *Math for All*

Given meaning through mathematics and learning psychologies, the equal sign in the curriculum works to organize identities and produce distinctions among people and objects in the world. This is made explicit in the relation between the styles of reasoning that organize knowledge of equality and children as particular kinds of people that intersect in the curriculum as rules for how to "know and do" equality as both a way of reasoning and a skill of determining equivalence and nonequivalence (Van de Walle, Karp, & Bay-Williams, 2010). In this assemblage, the equal sign inscribes a comparative logic about how to produce the individual who reasons about equality as sameness and inequality as difference.

Starting from contemporary research in teaching the equal sign, the discussion here has pointed to how the equal sign operates in the curriculum through models that give a fixed quality to knowledge and people. This was explored in the true/false problem-solving whereby what counts as math is validated through the use of numbers as telling the truth and inscribing a style of reasoning about equality. Embodied in that style of reasoning are specific sets of practices by which a child is to demonstrate learning. Importantly, the same principles that organize equality as a fact in the curriculum also manage the identification of children, ordered as the same or not.

The production of difference in teaching and learning equality becomes visible in how the reason of the equal sign and equality are used to represent and compare children's abilities to "think mathematically" (Carpenter, Franke, & Levi, 2003). The chapter titled "Equality" begins with a visual representation of children's responses to the question: "What number would you put in the box to make this a true number sentence? $8 + 4 = \square + 5$." Numerical inscriptions represent and organize children's answers, their grade levels, and the frequency of responses ranging from 7, 12, or 17 to 12 *and* 17. This translation and categorization of children's numerical responses into a statistical representation entails the use of numbers to render things (and people) in the world as fixed and possible to compare.

As a strategy for objectifying things in the world, numbers (and the notion of equivalence they embody) stand in to signify children's thinking. As such, they establish a fixed norm as the standard by which to compare children. The percentages are used as measurements to sort students' responses as deficient differences as compared to "most," "typical," or "common" (ibid, p. 1). The psychologies of learning and the rules of mathematical thinking contribute to this organization, granting equality and equivalence as a technical practice of learning mathematics as well as a cultural practice in providing ways of establishing equivalence and nonequivalence among children.

The comparative way of thinking about children is clear in how "children with learning disabilities" are represented as individuals with "severe disabilities with sequencing, decoding text, long term working memory, central auditory processing disorders, attentional deficits with and without hyperactivity" (Beatty & Moss, 2007, p. 30). "These" children do not simply exist but are fabricated with particular qualities, distinct from those who appear to "think mathematically" or as "typically developing children" (ibid, p. 35). This places the child associated with certain characteristics on a continuum of normativity, whereby the logic of equality reinserts difference as inequality. The commonsense descriptions of children's knowledge about equality seem to identify distinct kinds of people, one being desirable and the other being the difference that stands in the way of *all* children knowing mathematics.

Embedded in the notion of *all* is a representational logic that entails identification or making identical. Resolving and reducing the world into identifiable categories represent particular children as certain kinds of people. Along this distinction, rules about remediation and management of the seemingly problematic groups are organized by a similar logic that sets the task in school mathematics to fix inequality by identifying difference and making same. Although seeming to operate in distinct registers, notions of social and mathematical equality require a representation of stabilized identities and distinctions to determine and produce equivalence. Yet the normalizing process of reducing *all* to a common measure represents difference as deviance and inequality, justified through the reason of making equality.

Rethinking the Equal Sign, Reframing the Questions

This book aims to open a space to explore the cultural politics of school math. It draws attention to how the use of the equal sign in elementary school math is linked to cultural rules of equality, equivalence, and difference. The analysis of principles that order equality pedagogy in math reforms aimed at *math for all* entails looking at the curriculum in new ways and asking new questions about what mathematics does in the school. Within overlapping historical, cultural, and political registers, the book examines how the curriculum aimed at *math for all* organizes kinds of people through math education by asking:

- How does the use of the equal sign in the elementary curriculum embody rules for inclusion, equivalence, and difference?
- How are these principles inscribed in the curriculum to give identities to and organize differences among children? How do these change over time?

- In what ways do the norms for inclusion and equality embedded in the curriculum historically (re)define terms of exclusion and inequality?

Situating the problem in the ways of reasoning about equality, access, and inclusion goes against the grain of contemporary studies in curriculum reform and math education research. It goes against the grain in that it does not assume the issues and proposed solutions for educational change rest in the children or teachers. Instead, the problem is situated around how efforts to make children into mathematical kinds of people are organized by a logic that reinserts differences as the inequality the reforms aim to upend.

In contemporary studies, there is a paradox in the assumption that mathematics education is simultaneously considered a "great equalizer" (Musca & Menendez, 1988) and a "gatekeeper" to success and opportunity (Witzel & Riccomini, 2011). This framework is unquestioned yet places students on one side of the gate who seem to achieve and succeed according to the standard measures, whereas "others" appear to lag behind and represent the inequalities reforms seek to address. The questions of reform and research efforts, then, are aimed at determining how the gap in achievement can be closed, how more children can learn mathematics more efficiently, what the best practices used by effective math teachers are, and how the differences in children can be understood to create more equitable teaching and eradicate social inequalities.

Embedded in these approaches to school math research and reforms is a tradition of *critical mathematics* or *mathematics for social justice* (Frankenstein, 1987; Gutstein, 2006; Gutstein & Peterson, 2006; Skovsmose, 1994). Focused on the solutions that mathematics education can provide to the problem of social inequalities and exclusions, much of the scholarship aims to offer methods for progress—construed as better teaching, more efficient learning, a more inclusive classroom, or an improved multicultural and responsive curriculum for all children. With an impulse for inclusion to move society toward equality and prosperity for *all*, these studies and questions are necessary and important to school reform, but they are not sufficient.

Related to theories of critical pedagogy, critical mathematics and mathematics for social justice can broadly be seen as situated within an emancipatory pedagogy—that is, a pedagogy that is expected to free students from an otherwise oppressive and presumably hidden curriculum. The critical "problem-posing" approach to (mathematics) education is understood in opposition to an oppressive "banking method" (Freire, 1970). The assumption that the critical approaches aim to overturn is that a banking method—or filling the empty vessel—uses various psychological means to oppress the student and cut him or her off from knowledge coproduction.

The movement from a banking method of teaching to a problem-posing method is to bring the child from one psychological state to another—from astounded by the teacher to astounding, from empty to full, from being for someone else to being for oneself (Bingham & Biesta, 2010, p. 69). The critical method similarly applies a psychological "truth" about who children are and should be to (re)define how math education leads to either oppression or freedom.

Both a critical mathematics and mathematics for social justice aim to create the conditions for those who are not free to become free—by producing a psychological order where children are seen and see themselves as free. The image of the child as a math learner is a psychological one that progresses developmentally toward emancipation to live and learn in ways that are presumed to be free within a notion of social in/justice.

This freedom and the psychological traits that define its modes of participation are to be seen in how the child demonstrates rational thinking, independence, creativity, motivation, and confidence. These norms of participation become ways of classifying children and seem to make visible in/equalities. Yet they are expressed as psychological distinctions that have already been decided on in advance to order and divide the children who are free from those who need to be made free.

Attention to questions of who is and is not free and how *all children* can be free within a critical framework is important. Although they do not consider how the very ways of thinking about children embodied in this "counter" curriculum are taken for granted as natural ways of seeing, thinking about, and acting upon them. The dangers embodied in the approaches to critical mathematics can be provoked by the question: "Just what sort of human being is being assumed by the pedagogies that would otherwise claim to emancipate the human being?" (ibid, p. 63). Why the divisions are there in the first place, who gets to be seen as a math kind of person, what makes certain traits reasonable as ways to describe children and their learning, and how they are rationalized as necessary for inclusion and belonging have yet to be questioned.

This book aims to understand the limits of how the curriculum and its subjects have been formed. It does so by exploring how the logic of equality and inclusion are at play in cultural production of inequalities and exclusions in the math curriculum. Exploring the paradox is not merely to critique the present but to understand the rules that intern and enclose the possibilities for how to see, think about, and act upon the making of *math for all*.

Studying the school math curriculum as a site where children's identities are culturally and historically produced is to call into question how the image of the child as part of *math for all* has acted politically to draw boundaries that police who children are, who they should become, and who is not that child. This politics intersects with and becomes visible through the consistent application of psychological rules to the improvement of education that seems to create an equitable and inclusive social order through the organization of the child and the curriculum.

Within a larger narrative about learning mathematics in a democratic society, the child that is seen to be part of *math for all* is expected to embody certain psychological traits as the indicators of success that are culturally coded and inscribe rules for being seen as motivated, creative, confident, flexible, independent—and the list goes on. Yet the very ways of thinking about the equivalences that describe and bind the individual to the collective all as a form of equality and equity are unquestioned—even when they continuously produce "others" as different and excluded.

Notes

1. In reference to the Equity Principle for School Mathematics, the disregard of how differences and their production impact the reform of mathematics education is challenged throughout this book. Questioning how the reforms seek to shape access, opportunity, and achievement—irrespective of the presumed differences that characterize students—this study carefully works to understand how access and opportunity have been historically shaped with regard to the construction of differences that seem individual yet are culturally produced to make some groups representative of the standard and others as not meeting the standard.
2. Throughout, the discussion of the learning sciences refers to the scientific languages and ways of knowing that have come to characterize, explain, and shape learning as an activity organized by mental and social processes. Looking more specifically at how cognitive psychology emerged as a formal school of thought post-World War II, Chapter 3 discusses how a shift in focus on the mind and its internal processes became seemingly objective ways to explain and characterize learning. The focus throughout the text is in how the tools and principles of learning sciences intersect with the mathematical content to translate what students are to learn and shape how they are to learn it in ways that have implications beyond the mathematics.

References

Beatty, R. & Moss, J. (2007). Teaching the meaning of the equal sign to children with disabilities: Moving from concrete to abstractions. In Martin, G.W., Strutchens, M.E., & Elliott, P.C. (Eds.), *The learning of mathematics: Sixty-ninth yearbook*. Reston, VA: NCTM. pp. 27–42.

Bingham, C. & Biesta, G. (2010). *Jacques Rancière: Education, truth, emancipation*. New York, NY: Continuum.

Bloch, M.N. (1987). Becoming scientific and professional: An historical perspective on the aims and effects of early education. In Popkewitz, T.S. (Ed.), *The formation of school subjects: The struggle for creating an American institution*. New York, NY: The Falmer Press. pp. 25–62.

Carpenter, T.P., Franke, M.L., & Levi, L. (2003). *Thinking mathematically*. Portsmouth, NH: Heinemann.

Choike, J.R. (1996). *How it all adds up: Creating an agenda for all children*. New York, NY: College Entrance Examination Board.

Danziger, K. (1997). *Naming the mind: How psychology found its language*. London: Sage Publications Ltd.

Falkner, K.P., Levi, L., & Carpenter, T.P. (1999). Children's understanding of equality: A foundation for algebra. *Teaching Children Mathematics*, 6(1), 232–236.

Fennema, E. & Meyer, M.R. (1989). Gender, equity, and mathematics. In Secada, W.G. (Ed.), *Equity in education*. London: Falmer Press. pp. 146–157.

Foucault, M. (1991). Governmentality. In Burchell, G., Gordon, C., & Miller, P. (Eds.), *The Foucault effect: Studies in governmentality: With two lectures by and an interview with Michel Foucault*. Chicago, IL: University of Chicago Press. pp. 87–104.

Frankenstein, M. (1987). Critical mathematics education: An application of Paulo Freire's epistemology. In Shor, I. (Ed.), *Freire for the classroom: A sourcebook for liberatory teaching*. Portsmouth, NH: Boynton/Cook. pp. 180–210.

Frankenstein, M. (1995). Class in the world outside the class. In Secada, W.G., Fennema, E., & Adajian, L.B. (Eds.), *New directions for equity in mathematics education*. New York, NY: Cambridge University Press. pp. 165–190.

Freire, P. (1970). *Pedagogy of the oppressed*. New York, NY: Continuum.

Gutstein, E. (2006). *Reading and writing the world with mathematics: Toward a pedagogy for social justice*. New York, NY: Routledge.

Gutstein, E. & Peterson, B. (2006). *Rethinking mathematics: Teaching social justice by the numbers*. Milwaukee, WI: Rethinking Schools, Ltd.

Hacking, I. (2002). Inaugural lecture: Chair of philosophy and history of scientific concepts at the College de France. *Economy and Society, 31*(1), 1–14.

Musca, T. (Producer) & Menendez, R. (Director). (1988). *Stand and deliver* [Motion picture]. United States: Warner Bros.

National Council of Teachers of Mathematics. (1980). *An agenda for action: Recommendations for school mathematics of the 1980s*. Reston, VA: NCTM.

National Council of Teachers of Mathematics. (1989). *Curriculum and evaluation standards for school mathematics*. Reston, VA: NCTM.

National Council of Teachers of Mathematics. (2000). *Principles and standards for school mathematics*. Reston, VA: NCTM.

National Governors Association Center for Best Practices (NGA) & Council of Chief State School Officers (CCSSO). (2012). *Common core standards for mathematics*. Washington, DC: Authors.

Oakes, J. (1985). *Keeping track: How schools structure inequality*. New Haven, CT: Yale University Press.

Olkowski, D. (1999). *Gilles Deleuze and the ruin of representation*. Los Angeles, CA: University of California Press.

Patriarca, S. (1996). *Numbers and nationhood: Writing statistics in nineteenth-century Italy*. Cambridge: Cambridge University Press.

Phillips, C.J. (2015). *The new math: A political history*. Chicago, IL: University of Chicago Press.

Piaget, J. (1950). *The psychology of intelligence*. New York, NY: Harcourt, Brace & Co., Inc.

Poovey, M. (1998). *A history of the modern fact: Problems of knowledge in the sciences of wealth and society*. Chicago, IL: The University of Chicago Press.

Popkewitz, T.S. (2004). The alchemy of the mathematics curriculum: Inscriptions and the fabrication of the child. *American Educational Journal, 41*(4), 3–34.

Popkewitz, T.S. (2008). *Cosmopolitanism and the age of school reform: Science, education and making society by making the child*. New York, NY: Routledge.

Rancière, J. (2010). On ignorant schoolmasters. In Bingham, C. & Biesta, G. (Eds.), *Jacques Rancière: Education, truth, emancipation*. New York, NY: Continuum. pp. 1–24.

Scott, J. (1992). Experience. In Butler, J. & Scott, J. (Eds.), *Feminists theorize the political*. New York, NY: Routledge. pp. 22–40.

Secada, W.G. (1995). Social and critical dimensions for equity. In Secada, W.G. (Ed.), *Equity in education*. London: Falmer Press. pp. 146–164.

Skovsmose, O. (1994). Towards a critical mathematics education. *Educational Studies in Mathematics, 27*, 35–57.

Van de Walle, J., Karp, K., & Bay-Williams, J. (2010). *Elementary and middle school mathematics: Teaching developmentally*. Boston, MA: Allyn & Bacon.

Walkerdine, V. (1988). *The mastery of reason: Cognitive development and the production of rationality*. New York, NY: Routledge.

Witzel, B. & Riccomini, P. (2011). *Solving equations: An algebra intervention*. Boston, MA: Pearson.

3 Postwar Planning, Reforming Mathematics, and the Cultural (Re)Production of Children

School math reforms constantly waver between questions of *what* children should learn and *how*. Within this frame, making *math for all* students has become a common way of thinking about how to organize schooling in the United States. Yet the contemporary moves to improve math education are not about mathematics per se. As discussed in the previous chapter, math education in the United States has been tied to ways of thinking about how to organize society through notions of accessibility, inclusion, and equal opportunity. The purposes of school math in the United States have also been linked to ideas of progress in terms of technological, economic, and scientific advances. Yet despite debates over *what* children should learn and *how*, it has historically come to make sense that the future of society depends upon all students learning mathematics.

In the movement between concerns about the child and the mathematics, the pendulum of reform takes as granted that to change the effectiveness of math instruction and its impact on society, one must remake the curriculum and the child learning it. Elaborated in the previous chapters, children and their learning in schools have been presumed as the site of social improvement and change. Rather than taking this phenomenon for granted, this chapter raises questions about the relationship between elementary mathematics education and how it has become implicated in reform projects to improve society. It asks how it historically became possible to relate children's math learning to ideals of citizenship, social equality, and notions of social progress. In the end, it asks about how the content of the math curriculum and moves to make *math for all* intersect with cultural rules and narratives about who children are and who they are supposed to become through mathematics.

This chapter examines the historical ties that bind children's mathematics learning to living as certain kinds of people in what was thought of as a democratic and progressive society after World War II in the United States. This relationship and *math for all* emerged at the intersection of three historical conditions that will be explored here: 1) the promise of math and science as useful for engineering solutions to social problems, 2) the reemergence of the social sciences as a way to organize a rational individual and society, and 3) the role of cognitive psychology in knowing and imagining the child as integral to society's development. At the crossroads of these three historical ways of reasoning about children, mathematics, and planning society, there emerged a new way of thinking about how to reshape children's math learning to reform society.

The aim here is to raise questions about the relationship between elementary mathematics education as it has become implicated in reform projects to improve society. It asks how it historically became possible to relate children's math learning to ideals of citizenship, social equality, and notions of social progress. In the end, it asks about how the content of the math curriculum and moves to make *math for all* intersect with cultural rules and narratives about who children are and who they are supposed to become through mathematics.

These questions shift the focus in math education research from prescribing how to improve the child or the curricular content to examining how the elementary math curriculum in the United States orders historical and contemporary ways of thinking about school mathematics as a tool to shape children and society. This chapter discusses how movements toward *math for all* emerged in the post-World War II United States. It explores the historical conditions in which *math for all* became reasonable within shifts in the relationships among mathematics, children, and the production of what was envisioned as a democratic, progressive, and inclusive society. In this shift, *math for all* reforms can be understood within the social goals of making children into certain kinds of people.

Math for All as a Historical Event: A Cultural History of the Present

Although a commonsense idea now, it has not always been possible to think that all children could or should learn math. Historically, mathematics was seen as a form of thought too abstract and therefore was considered to be valueless, impractical, and even dangerous—particularly for young children (Langer, 1931). Since that time in the history of U.S. school mathematics, the place and value given to mathematics in the elementary school has shifted. The appearance of the Common Core Standards for Mathematical Practice in the United States in the early 21st century highlights the ways in which mathematics is no longer considered to be dangerous for children. As part of an effort to provide what has been conceived as coherent and consistent instruction in math across the nation, the Common Core Standards were framed around the goal that math is to be seen by students in kindergarten through high school as sensible, useful, and worthwhile (NGA & CCSSO, 2012). This historical shift in the value given to all students learning math—particularly young children—signals a historical event and raises the question of how it has become reasonable to think that *all children can and should learn mathematics*.

As a moment in which it appeared to make sense that math is for all, including children, the emergence of *math for all* is significant. This chapter aims to understand how elementary mathematics education and its reforms have been positioned as integral to a child's schooling. But this is not just about school or mathematics. Underlying the *math for all* movement is a way of thinking about the relationships among mathematics, individual development, social progress, and the identity that seems to have traveled internationally and within the United States as emblematic of its position as a democratic and major industrial nation.

Discussed in the introduction, the book approaches *math for all* as a "cultural history of the present" (Popkewitz, Franklin, & Pereyra, 2001). The chapters move back and forth between the past and present to understand the rules that organize *math for all* reforms in the elementary math curriculum in the United States. Studying this shift toward *math for all* is to understand how reforms, both historically and in the present, operate to define what that social equality looks like while classifying and producing differences in children that are given as the reason for inequality.

To see *math for all* as an event and an invention opens up a space to examine the role that was given to math in (re)making the United States after World War II, which will be further elaborated in the coming sections. At the intersection of postwar ways of reasoning about children, mathematics, and planning society, there emerged a new way of thinking about how to reshape children's math learning to reform the nation. As a historical event, this emergence of *math for all* is understood given the historical conditions wherein children's mathematics became part of the formula for social progress toward equality, access, and opportunity.

Math for All as a Tool for Making the Modern Life

It was not until after World War II in the United States that it became reasonable to believe that *all children's* ability and achievement in mathematics would improve society. The belief in *math for all* appeared as a technical solution to social problems and part of the formula for national progress after the war. This imperative toward social change was expressed in national projects that sought a Great Society, waged a War on Poverty, fought for civil and women's rights, and generally defined progress in terms of technological advancements, social inclusion, and economic growth for all. The goals to eliminate economic, racial, and gender injustices through these projects were tied to new ways of thinking about planning society. Embedded in efforts to bring the United States "within the framework of freedom, to increase the living standards of the majority of people and at the same time maintain or raise cultural levels" was a new logic that society and its future were not the outcome of strict ideologies but could be systematically planned (Bell, 1962).

As a shift from previous ways of reasoning about social change as a factor of ideologies and predetermined beliefs, the problems in U.S. society after World War II were viewed as technical, whereby their solutions could be engineered. The social movements mentioned earlier were not just ideas about how to improve a standard of living and raise cultural levels but were tied to new practices of using scientific research and mathematical modeling to plan society.

During and after World War II and into the Cold War period in the United States, mathematics assumed a new cultural relevance. Taken as a "tool of modern life" (College Entrance Examination Board [CEEB], 1959), mathematics became normalized as a useful mechanism for representing and organizing society. The presumed utility of mathematics was not about mathematics per se but about how it was thought to stand in as a form of rationality that could be technically applied for modeling, calculating, and predicting the unknown. This logic ascribed to math is expressed in how it would be seen as the "Queen and Servant of the sciences" (School Mathematics Study Group, 1966a, n.p.) during the new math reforms and imagined as integral what appeared as the scientific production of technological, economic, and social progress.

This hope invested in the rationality of scientific and mathematical thinking is visible in how mathematical models and calculations were supposed to produce solutions to technical problems. The Electronic Numerical Integrator and Computer (ENIAC) that appeared during the war embodied a way of thinking about the practicality of mathematics for effectively solving problems. Its main function was to decode algebraic expressions to solve for unknown variables of wind resistance, distance, and speed, thereby calculating artillery-firing tables for the U.S. Army. The application of mathematics in its service to physical sciences was to plan the future of national security via technological progress.

This use of mathematics for modeling and planning the future was embedded in what was considered a "crisis" caused by the "breath-taking movement into the new technological era" (CEEB, 1959). In the midst of a perceived crisis, it would come to make sense "how urgently the general student needs mathematical instruction which will keep pace with the rapidly increasing technological demands of our society" (Stone, 1957, p. 73). This presumed need of math instruction for a "general student" was tethered to the belief that math is useful in manufacturing and managing the future by planning the kind of people who can think logically with mathematics as a proxy for rational and organized thought. Within this historical turn, the logic that *all children* should learn mathematics was not merely about mathematics but embodied a way of reasoning about how to use mathematics as a technology for solving society's problems.

Signifying a shift in reasoning during this postwar moment, the logic given to mathematics as a tool for planning a "modern life" was not only deemed useful in solving technological issues—like computer coding for firing missiles—but social ones as well. Provided the presumably neutral tool of mathematics, it was assumed that "measurement in psychology and physics are in no sense different. Physicists can measure when they can find the operations by which they may meet the necessary criteria; psychologists can do the same" (Reese, 1943, p. 49). The "scientific" practices of measurement and calculation in psychology and physics were assumed to be equivalent through the rationality of math as the technical reason of science. This use of mathematics to measure human behavior and model social phenomena articulates how the promise of math as a tool for making modern life intersected with the (re)emergence of the social sciences as a mechanism for planning how people were to live this life.

Planning Society and People Through the Social Sciences

Following World War II in the United States, science (and mathematics in its service) became normalized as a form of rationality presumed useful for measuring, organizing, and modeling what was envisioned as a modern life. The notion of rationality embodied in the use of mathematics—planning, modeling, and predicting—was integral to defining what that modern life was and who was to live it. More specifically, this new cultural value granted to mathematics was embedded in a new social logic that organized who the postwar citizen was and should be.

On the surface, the imperative for more people to learn mathematics was situated in questions about technological advancements. Yet, beneath that was a way of thinking about how the use of mathematics in planning society was also tied to notions of planning a certain kind of person who could think and act in particular ways. After the war there arose a pessimism toward and loss of faith in human rationality—a capacity of thought upon which a modern democracy seemed to depend (Heyck, 2012, p. 100). The fear that people could not be trusted to make rational decisions, particularly after Nazism and Fascism, was coupled with a belief that "the rational individual is, and must be, an organized and institutionalized individual" (Simon, 1961, quoted in Heyck, 2012). Thus emerged the common sense to govern the processes of thought and action associated with rationality so that it could be inscribed in humans, even if they were otherwise deemed irrational. The institution and the rules for making this so-called rational person arose out of the reconfiguration of the postwar social sciences in ways that created new possibilities for governing how one should live a modern life.

The institutionalized belief in and study of what was thought of as a *rational citizen* espoused the hope that democracy—in spite of irrationality, fallibility, and authoritarianism—could be maintained and recognized through a particular set of behaviors and traits thought to make up that kind of person (and thereby a rational and democratic collective).

In the face of doubts about human reason, postwar social science made visible and knowable a rational kind of person to plan society. A postwar reformulation of the social sciences in the United States re-signified them as "the policy sciences" (Lasswell & Lerner, 1951), a characteristic of their 19th-century formation but given a more public expression in the postwar years. This new configuration meant that they were publicly positioned to study society as a complex system, identify its problems, and provide the rules and interventions that would (re)organize what was seen as an inter-related, collective society.

Changes within postwar social sciences were bound to the belief that the social operated as a biological system of interacting variables. In the mid-20th century United States, this living property of systems ascribed to the social signaled a turn in social and human sciences "away from the atom-ized individual and towards collectives as the source of social stability . . . individual parts were important but their robustness or fallibility could not be understood without examining the larger environments that surrounded them" (Jones-Imhotep, 2012, p. 179). Studying society in a seemingly sys-tematic way relied upon an approach to modeling the social after an organ-ism to plan for equilibrium as its most efficient state of operation. This way of seeing the social—not as a technical machine but as a living entity—allowed for envisioning the manufacturing of change but at the same time inscribed a sense of certainty in how the system functioned.

The view of the social as a living, biological system was tied to the belief that an individual and the collective group were intricately bound. This way of seeing society as an interrelated network, rather than a hierarchy between individual and collective, was a historical shift that distinguished postwar social sciences from its turn-of-the-century iterations (Cravens, 2012, p. 120). With this new way of seeing and reducing the relationship between the individual and society came "a new emphasis on quantification, or even mathematization" of people, their presumed problems, and prescriptions for solutions (ibid, p. 121). Promising a sort of objectivity and precision, studies of human behaviors, attitudes, and beliefs could now incorporate an F-test, chi-square, or multiple regression analysis with human nature as the measurable variables in formulas for change and progress.

With shifts in how the social could be studied, measured, and analyzed, the social sciences brought new objects of study into the purview and with them new traits by which to redefine who would be the progressive and democratic citizen. This was coupled with a movement toward social proj-ects that embodied the hope that any perceived crises could be managed through science and the making of rational kinds of people.

The use of mathematics in the social sciences uplifted and validated their place in the academy in line with positivist and empirical approaches to science. Even more, the new mathematical methods for modeling human behaviors made it so the study of an individual could represent any indi-vidual and, therefore, a group (ibid). This conflation between the individual and the group in the reimagination of a democratic social order coincided with a new way of measuring and organizing human behaviors. Connected to broadened social goals of understanding society by understanding the group and individual, this measurement was not merely about doing mathematics. Rather, its practices of modeling and quantification were embedded in the value that was given to math, science, and technology as useful tools for planning the future of the postwar nation.

As a cultural practice, the use of mathematical modeling in the social sciences was deemed useful in planning how each individual could become a certain kind of democratic and rational person who would live the modern life. This construction of a so-called democratic identity in the United States after World War II and into the Cold War can be seen in *The Authoritarian Personality* (Adorno et al., 1950). Through the measurement and calculation of nine personality traits, which were thought to collectively represent one's propensity toward authoritarian behaviors, individuals were calculated along an F scale. The study was seen as an empirical examination of broad and coherent patterns of the personalities of individuals who would be identified as fascists (thus, F scale).

This group of individuals characterized by what was deemed an "authoritarian personality" was measured in relation to characteristics defining a "democratic personality." The personality tests expressed a way of thinking about how to create a sense of unity within society through the formation of a cohesive collective. In this formation, the "democratic personality" was not merely a psychological distinction. It was a cultural one that gave meaning to what it meant to live as a rational citizen, further elaborated in the next section.

As historical conditions whereby *math for all* became possible to think, the promise of math and science converged with the shifting focus and methods of the social sciences to produce the reason that the modern life could in fact be made through the fabrication of a rationality that was embodied in mathematics through modeling, predicting, and organizing for the future. The belief in a democratic and rational kind of person who would live and make the modern life were integral to positioning social sciences and math with a role in planning people as a way of planning the future of society. This logic about mathematics and ordering society would intersect with new ways of thinking about how to understand, model, and develop what was perceived as a rationally thinking person through the cognitive sciences and psychological studies of the child.

Cognitive Sciences and Making a "Rational" Social Order

This discussion pointed to the historical reconfiguration of how, through social science research, mathematics became integrated into social policy by measuring, organizing, and modeling a norm of rationality in "democratic" kinds of people. Here the role of cognitive psychology is examined in terms of how it functioned to define characteristics of the mind that would be inscribed in the child through mathematics as a tool to develop rational, democratic citizens.

As a newly emerging field working to legitimize the scientific study of the mind and its inner workings, cognitive science in the mid-20th century had to overturn the predominant narrative that the mind did not exist as a matter of significance in understanding human behaviors, attitudes, reason, or motivations. In fact,

> from the 1920s to the 1960s, behaviorist psychology held sway as the scientific approach to human nature. With the argument that science required objective observation of measurable phenomena, and that mental phenomena being immeasurable, either should be left to the philosophers or, like the soul, simply did not exist, behaviorists had managed to gain control of the scientific and rigorous end of psychology.
> (Cohen-Cole, 2014, pp. 6–7)

But if individuals could be organized to behave rationally, despite a presumably fallible and irrational nature, then the belief that humans act solely in response to external conditions would not be sufficient for explaining human nature. At this historical intersection of an appeal to the human as a rational decision maker and a compulsion to study the mind with scientific measures and quantification, cognitive psychology began to take hold as an explanatory force for social facts and human nature. In its postwar appearance, the cognitive view was (re)visioned as the predominant way to understand, explain, and organize human behaviors.

With practices aimed at solving social issues, the research and development of cognitive psychology in the postwar United States was to provide models of behavior and thought processes presumed to represent people and their social problems, identify causes, and manage solutions. Within this frame, psychological research was "perceived as a vehicle that will assist in bringing about the American Creed of equality, fair play, and minimal group conflict" (Darley, 1952, p. 719). Embodied in this statement is the common sense that society could be made more coherent and equal through psychology's role in defining the kinds of people who were seen as rational, fair, and democratic.

Cognition and a rationally working mind became objects of scientific study in ways that historically had not been possible before. This shift assembled with new ways of reasoning about who is and should be a rational and democratic citizen. This kind of person was thought of as an active and autonomous decision maker, not a passive bystander who behaved in rigid ways with predetermined responses to their environment (Cohen-Cole, 2014). As distinct from the behaviorist perspective, the cognitive view promoted a more infinite range and freedom of conduct through this model of rationality. To this end, "cognitive scientists not only epitomized the democratic character, but their account of humanity was more attractive. To accept their scientific vision was to find that being quintessentially American was one and the same as being human" (ibid, p. 6). The psychological tools that were deployed in measuring and making the rational mind visible through cognitive science turned the valued mental virtues of rationality into civic virtues of a presumably "normal" American citizen.

At a time when the social interest was in eliminating a fear about the irrationality of people, it became sensible that the cognitive and social development of a society would depend upon the psychological development of a form of rationality. This development was understood through the processes and activities of the mind defined through cognitive psychology.

By identifying the kind of person who would "bring coherence to America's increasingly complex and diverse culture" (Cohen-Cole, 2009, p. 220), cognitive psychology organized the idea of the democratic citizen around notions of freedom of thought, reason, tolerance, diversity, and creativity (ibid, p. 226). Imagination, innovativeness, flexibility, autonomy and an "open-ego" were also given as characteristics of the kind of citizen a modern and progressive democratic society required (Doob, 1960; Gardner, 1963; Maslow, 1943). More than psychological traits, these cultural ideals were ascribed to individuals through the measurement of what were defined as civic behaviors and democratic attitudes.

Postwar cognitive sciences made further appeals to the public image of a rational kind of person by employing specialized sciences in defining what that mind and its inner functionality looked like. In doing so, the psychological expressions of cognition were associated with computer sciences. Cognitive psychologies used the tools, models, and language of computers and cybernetics to explain the mind and its functions. To give more public expression to a seemingly expert knowledge, the mind was envisioned as a sort of computer system or network that processed information and produced outputs.

Through the image of the computer and the language of the mind as a series of networks and programs, the cognitive sciences deemed it possible to infer about the human mental processes that were previously thought invisible. This can be seen in studies of human problem-solving and computational theory wherein the brain is defined as a logical machine (McCulloch & Pitts, 1943; Newell, Shaw, & Simon, 1958). Thought processes became the internal function of presumably universal human capacities that were granted as necessary for cognition and its visible output as learning in all people. This knowledge of the mind was not merely a technical knowledge—it was a cultural one expressed in the language of psychology whereby the networks and wires were soldered to mental traits and became the explanation for human behavior in terms of motivations, emotions, behaviors, and thoughts.

This new form of knowing the mind held purchase in defining the role that cognitive psychology could play in making a new form of rationality discernible and developing all people as the kind who appeared to think and act rationally. The computer model provided a seemingly universal mental mechanism that could model thinking processes that were already granted importance in defining human nature in postwar America (Cohen-Cole, 2014).

Given this apparent universality in how the mind functioned, concerns about shaping individual character through the development of rationality could then cut across variations in culture and age. That is, if the supposed rationality of a democratic citizen was something that must—and indeed could—be developed across human nature despite differences and irrationality, then the importance of the child would come to bear on organizing the sought-after democratic development of the nation.

Child Study and the Development of a Nation's Citizenry

Making democracy through the fabrication of a rational citizen after World War II was primarily seen as a psychological problem. In this way, then, the response was also a psychological one—with a new emphasis on the child as part of the solution. In the fabrication of the modern and democratic citizen, the child took on a new social importance.

Children's place in the development of society was expressed in the rise of child study and the belief that childrearing was integral to building the character of a society. In this historical moment, the key question for child psychologists was how the socialization of a child later impacted the personalities of the adult and the social order (Vicedo, 2012, p. 236). Raising the child as a function of building a future society gave rise to a growing significance of the field of child development and psychological concerns about childhood. The belief that children develop in natural and linear ways toward adult thought processes and patterns of behavior contributed to the idea that the child could play a part in rearing future society to be rational and independent as a democratic nation.

The link between a child's individual development and society can be seen in *Childhood and Society* (Erikson, 1950). This text articulates the importance that was placed on the theory that children develop continuously throughout life in a way that has an impact on society. In a way that suggested a blueprint for human psychological growth, the emotional growth of a child presumably mapped out the makings of a stable, mature, and psychologically evolved society.

The idea that children could be reared into rational citizens—defined through the principles of cognitive psychology—would become part of the common sense of technically planning the postwar "modern life." The plans were bound up in how "man must decide whether he can afford to continue the exploitation of *childhood as an arsenal of irrational fears*, or whether the relationship of adult and child, like other inequalities, can be raised to a position of partnership in a more reasonable order of things" (ibid, p. 47, my italics). Highlighting childhood anxiety, confusion, and fear as effects of biological, social, and psychological causes suggested that deciding about the place of the child entailed a decoupling of the association that had been made between children and a presumed innate irrationality as the locus of behavior. In line with "a more reasonable order of things," a new way of thinking about children emerged. Reimagining children developing along an evolutionary trajectory of rationality—rather than presumed irrational—would give them a role in producing and maintaining a democratic society.

Emerging as part of a solution to social problems, it seemed sensible that a rational, democratic, and modern kind of life could be made through mathematics. In this new configuration, math education found a place as a new sort of social policy that required knowing and developing the child as a rational kind of person. Expressed as a way to deal with "crucial problems" and "strengthen math curriculum," there was "no doubt that the key is to be found in a better understanding of psychology of the child" (Stone, 1957, p. 75). This way of thinking about math learning, children's development as rational people, and the future of the nation became part of the commonsense reasoning about reforming math as a way of planning people. Throughout the book, we will see how the reasoning that math is for all assembles with the tools of psychology as a pedagogical science to order the child's thoughts and actions in ways that historically (re)shape what it means to be part of the *all*.

To this point, this chapter has examined the interrelations among three historical conditions: 1) the promise of math as a tool for living the modern life, 2) the reemergence of the social sciences as a way to plan the rational individual and democratic society, and 3) the role of cognitive psychology in organizing that rationality in the child as integral to society's development. It has explored how shifting ways of thinking about the relationships among children, democratic society, and mathematics made it possible to think that math should and could be for all in the postwar United States. The next section explores how mathematics education for all became a technical way of producing the conditions whereby children could become rational citizens who could cultivate an inclusive and equitable democracy.

Equal Opportunities for All—a New Commonsense About (Math) Education

Historically, it was not new that the school would be implicated in solving social problems and socializing the public (Smeyers & Depaepe, 2008). What was new in the postwar period, however, were the forms that the problems and solutions took. In response to perceived crisis in the United States, mathematics education was positioned as part of the solution to the social problem of remaking and sustaining a democracy.

In this new configuration, appeals to reorganize and unify the elementary through the postsecondary math curriculum moved beyond the link between mathematics and modern technology. Desires to reform the curriculum were also imbued with values about what it meant to be educated and live as a citizen in a modern, democratic democracy. This was visible in the idea that the "wisest plan is to offer sound basic mathematical instruction for all" (Stone, 1957, p. 67). More than about mathematics, this new wisdom carried principles of social inclusion, expressed as an expansive and collective "all" that would become makers of the postwar democracy. Embedded with principles of equality and equity, the reformulation of math education inscribed commonsense thinking about how building math education for all could serve the democratic function of making schooling more equal and society more just.

This educational and social goal toward inclusion and equality was not limited to math. It was also discussed in general curriculum reforms. In a 1944 report titled, "Education for All American Youth," the Educational Policies Commission (EPC) claimed that when

> we write confidently and inclusively about education for *all* American youth, we mean just that. We meant that all youth, with their human similarities and their equally human differences, shall have educational services and opportunities suited to their personal needs and sufficient for the successful operation of a free and democratic society.
>
> (p. 17)

The call for *all* youth to have opportunities to live and learn through schooling was taken as the expression of an inclusive, free, and democratic society. On one hand, this inclusivity was a way to build unity and create a cohesive nation. On another, this call for education for all was also tied to a new emphasis on egalitarianism in the form of equality of opportunity—particularly with regard to school organization. As a new way of thinking about the social function of schools, equality of opportunity was related to the logic that schooling and all that it could offer was considered a right that should be granted to all (Bankston & Caldas, 2009). Naming this right, however, required identifying those who did not already have it.

With hopes for inclusion and equality tied to schooling, there emerged a public expression of concerns that school was not open to all, and in fact unequal opportunities existed in the form of facilities, resources, and materials (ibid, p. 101). These concerns were not simply about run-down buildings. They were about how the visible differences in schooling became a representation of inequality. These differences—in rights, resources, and opportunities among individuals—were and still are seen as the problem that stands in the way of achieving equality as a fundamental principle of democracy through education. Moves to make school more generally *for all* were reasoned about as a way of equalizing access to greater social, economic, and political opportunities. For "all" was a call to include those who had previously been excluded.

Given "equally human differences," in movement toward "education for all youth," the EPC considered "human similarities" to be evident in the social fact that "all American youth are citizens now" (EPC, 1944, p. 16). In spite of what otherwise seemed to divide all young people worthy of an education, they were all thought to be citizens with rights. With this new way of thinking about all youth as part of collective society with access to social opportunities came the belief that "all American youth have the capacity to think rationally; all need to develop this capacity, and with it, an appreciation of the significance of truth as arrived at by the rational process" (ibid). Although on the surface, distinctions of the human and citizen appear obvious, to be seen as either and therefore deserving of the equal rights embedded in the promise of a democratic and egalitarian society required living in particular ways. That is, with the right to education granted to all youth as human citizens came the responsibility to think and act as a rational kind of person.

Calls to remake the math curriculum during the postwar period were expanded to include young children as part of a collective all. Given the historical meaning given to rationality as a psychologically developed quality of being human (rather than a fixed state of mind), it became reasonable to think that a reformed math curriculum could and should be opened to include all as a way of making all youth develop as fully human, rational, citizens.

Whereas the curriculum reforms of mathematics had previously focused on secondary and postsecondary education, the moves to improve access and opportunities for all students to learn math shifted to include the young. Given a presumed continuity between secondary and postsecondary school, the trajectory of curriculum planning extended into the elementary school as a foundation for both. This would eventually set the stage for projects to develop the elementary math curriculum in ways that would aim to make children into citizens who would use mathematics as a tool for modern life. It also produced the cultural norms that would signify differences as the problem of reform.

Backed by a change in the value given to children's math and federal funding (through the National Defense and Education Act and the National Science Foundation), *math for all* took off as a sort of social policy. As a matter of social importance, math was about more than economic or technological advances. Reorganizing the math school curriculum from the elementary level up was also rationalized as a way to provide equal opportunities by developing all young people as citizens through the cultivation of the civic and mental virtues ascribed to the rational kind of person through mathematics. It would seem that the "crucial problems of math instruction" and of society could be solved through the child learning math and internalizing the qualities of the form of rationality ascribed to mathematics. This became a commonsense idea about reforms that would permeate projects to reorganize elementary math, both historically and in the present.

Drawing Boundaries to Examine In/Equality

This chapter has examined the bond that was historically formed between children's mathematics, social change, and planning people who are thought to live modern and democratic lives. This relationship significantly shifted the ways of thinking about the role of elementary math education from the mid-20th century on. At the historical emergence of the possibility of making *math for all*, the math curriculum has become a way to mobilize individuals, society, and the nation toward a future that not only seems more technologically savvy and secure but also one that appears more aligned with principles of democracy and equality through the fabrication of children as rationally thinking citizens.

The historical relationship between math education and planning children as future citizens of the nation is important in how the inscriptions of children's qualities and traits become visible and shift in curriculum reforms over time. Within this historical frame of post-World War II to the present, the following chapters look at three periods of math education reform to consider how elementary mathematics has become tied to notions of progress, development, and equality in ways that make children into certain kinds of people. In looking closely at the "new math" (late 1950s–early 1970s), "back to basics" (1970s–early 1980s), and "standards-based" (1980s–early 1990s) reforms,[1] each chapter considers historical shifts in reasoning about how math education functions in society by making the child into a mathematical kind who is thought to be part of the *all*.

Together the chapters highlight how the norms of identity that are to include all children into the *math for all* shift from ideals of citizenship, to distinctions of individuality, to standards of personhood in ways that come into the present as the commonsense expectations for how children should live and act as "equals." This way of thinking about how the math curriculum embodies social theories of equality requires a different look at the curriculum. Turning to these three historical markers is to consider how the curriculum—and the use of the equal sign in it—carries cultural practices that generate and maintain norms of identity, difference, and equivalence that operate in making a mathematics that is not quite for *all*.

Note

1. The time frames are borrowed from Klein (2003) and are useful in providing boundaries for primary and secondary document selection. However, they are not used to suggest that as one period "ended" another "began." Rather, they can be seen as overlapping historical trajectories that shape and are shaped by cultural practices and principles—giving meaning to the content of the curriculum and the subjectivities of the "learner." Attending to the ways in which there was not one singular reform movement, throughout the text the periods of reform are not capitalized or referred to as singular events but historical assemblages of cultural ways of reasoning.

References

Adorno, T.W., Frenkel-Brunswik, E., Levinson, D.J., & Sanford, R.N. (1950). *The authoritarian personality*. Oxford, England: Harpers.

Bankston, C.L., & Caldas, S.J. (2009). *Public education: America's civil religion, a social history*. New York, NY: Teachers College Press.

Bell, D. (1962). *The end of ideology: On the exhaustion of political ideas in the fifties*. New York, NY: Free Press.

Cohen-Cole, J. (2009). The creative American: War salons, social science, and the cure for modern society. *Isis, 100*, 219–262.

Cohen-Cole, J. (2014). *The open mind: Cold war politics and the sciences of human nature*. Chicago, IL: The University of Chicago Press.

College Entrance Examination Board. (1959). Program for college preparatory mathematics. In Bidwell, J.K. & Clason, R.G. (Eds.). (1970). *Readings in the history of mathematics education*. Washington, DC: National Council of Teachers of Mathematics. pp. 664–706.

Cravens, H. (2012). Column right, march! Nationalism, scientific positivism, and the conservative turn of the American social sciences in the cold war era. In Solovey, M. & Cravens, H. (Eds.), *Knowledge production, liberal democracy, and human nature*. New York, NY: Palgrave Macmillan. pp. 117–136.

Darley, J. (1952). Contract support of research in psychology. *American Psychologist, 7*, 719.

Doob, L.W. (1960). *Becoming more civilized: A psychological explanation*. New Haven, CT: Yale University Press.

Educational Policies Commission (EPC). (1944). *Education for all American youth*. Washington, DC: The national education association of the United States.

Erikson, E.H. (1950). *Childhood and Society*. New York, NY: Norton.

Gardner, J.W. (1963). *Self renewal: The individual and the innovative society*. New York, NY: Harper & Row.

Heyck, H. (2012). Producing reason. In Solovey, M. & Cravens, H. (Eds.), *Knowledge production, liberal democracy, and human nature*. New York, NY: Palgrave Macmillan. pp. 99–116.

Jones-Imhotep, E. (2012). Maintaining humans. In Solovey, M. & Cravens, H. (Eds.), *Knowledge production, liberal democracy, and human nature*. New York, NY: Palgrave Macmillan. pp. 175–195.

Klein, D. (2003). A brief history of American K–12 mathematics education in the 20th-century. In Royer, J. (Ed.), *Mathematical cognition*. Charlotte, NC: Information Age Publishing. pp. 175–259.

Langer, S.K. (1931). Algebra and the development of reason. *The Mathematics Teacher, 24*(5), 285–297.

Lasswell, H.D. & Lerner, D. (Eds.). (1951). *The policy sciences*. Palo Alto, CA: Stanford University Press.

Maslow, A. (1943). A theory of human motivation. *Psychological Review, 50*, 370–396.

McCulloch, W.S. & Pitts, W.H. (1943). A logical calculus of the ideas immanent in nervous activity. *Bulletin of Mathematical Biophysics, 7*, 115–133.

National Governors Association Center for Best Practices (NGA) & Council of Chief State School Officers (CCSSO). (2012). *Common core standards for mathematics*. Washington, DC: Authors.

Newell, A., Shaw, J.C., & Simon, H.A. (1958). Elements of a theory of human problem solving. *Psychological Review, 65*(3), 151–166.

Popkewitz, T.S., Franklin, B., & Pereyra, M. (2001). *Cultural history and education: Critical essays on knowledge and schooling*. New York, NY: RoutledgeFalmer.

Reese, T.W. (1943). The application of the theory of physical measurement to the measurement of psychological magnitudes, with three experimental examples. *Psychological Monographs*, 55, 1–89.

School Mathematics Study Group. (1966a). Mathematics for the elementary school: Book 1, Part 1. Teacher's commentary.

Simon, H.A. (1961). *Administrative behavior*. 2nd Edition. New York, NY: Macmillan.

Smeyers, P. & Depaepe, M. (Eds.). (2008). *Educational research: The educationalization of social problems*. New York, NY: Springer.

Stone, M.H. (1957). Some crucial problems of mathematical instruction in the United States. *The School Review*, 65(1), 64–77.

Vicedo, M. (2012). Cold war emotions: Mother love and the war over human natures. In Solovey, M. & Cravens, H. (Eds.), *Knowledge production, liberal democracy, and human nature*. New York, NY: Palgrave Macmillan. pp. 233–249.

4 Creating the Great Society

Making New Math, the Mathematical Citizen, and the Problem of Disadvantage

In popular histories of math education, the launching of the Sputnik satellite by the Soviet Union presumably sparked an urgent overhaul to the math and science curriculum in the United States (Dow, 1999). This historical explanation illuminates how, on the surface, the changes in math education were linked to concerns of anti-intellectualism, incompetence, and national security. Although Sputnik has become a convenient event in the history of math education, the changing role of mathematics education in the United States was not simply caused by a satellite or the threat of Cold War. Beneath the surface, the making of a new math curriculum was tied to a political and cultural narrative about making a new kind of nation in the post-World War II period.

With math taking on a new value in the remaking of the postwar nation, math education became integral to making future democratic citizens. During this time, elementary school math became a key focus of large-scale curriculum development reforms. These relationships among children, mathematics, and the nation became visible in the new curriculum to create new ways of characterizing children as intelligent and rational kinds of citizens. To further examine how the traits of citizenship were embedded in new math reforms, this chapter digs into how the elementary math curriculum inscribed the child as a future citizen of the modern nation.

Expressed in the language of psychology, this ordering of identity and equivalence made it reasonable to think that all students at all levels could and should learn mathematics. But the common sense of all students is a fabrication. *All children* are given materiality through characteristics given in the math curriculum, making up children as certain kinds of people (Hacking, 2002). As a way to understand the cultural production of identity and difference that represent who the child is and should be as rational, democratic, and intelligent kinds of citizens, this chapter highlights how mathematics was translated through principles of psychology to remake the math curriculum, fabricate a new kind of citizen, and generate principles about inclusion and difference.

Arguing that the math children were supposed to learn in schools had little to do with mathematics, it attends to how the curriculum embodied efforts to make a democratic and inclusive society by providing cultural rules for how children should think and act as citizens. To trace the logic of equality as a way of reasoning about and establishing what is equal, equivalent, and different, the chapter examines the use of the equal sign (=) in the new curriculum.

On one level, the equal sign is taken as foundational content of what elementary students should learn in reforms developed by the School Mathematics Study Group (SMSG). One another, the chapter explores the use of the equal sign as a cultural practice that embodies rules for how to learn that organize children's identities as intelligent kinds of citizens—characterized as creative, flexible, independent, and abstract thinkers. The analysis investigates how the identity given to children's mathematics and norms of participation as citizens defined the terms by which to classify the child who appeared as disadvantaged and stood as a reason for inequality. It argues how seemingly distinct mathematical and cultural notions of in/equality intersect to organize a new mathematics education—for all students at all levels—while simultaneously excluding the child marked as different.

Planning the Modern Life and a New Curriculum for All Students

After World War II in the United States mathematics was considered a "tool of modern life" (College Entrance Examination Board [CEEB], 1959). One only need to look to the introduction to a teacher textbook created by the SMSG to see this intersection between learning math in schools and living a modern kind of life:

> The increasing contribution of mathematics to the culture of the modern world, as well as its importance as a vital part of the scientific and humanistic education, has made it essential that the mathematics in our schools be both well selected and well taught—at all levels, from the kindergarten through graduate school.
>
> (SMSG, 1966a, n.p.)

At all levels—including the elementary school—math was emerging as part of education's vital and cultural importance. Yet this systematic approach to planning a "well selected and well taught" curriculum was not about mathematics alone. The movement toward a new curriculum, particularly in mathematics and sciences, was aligned with the hope that change would "lead the way toward inspiring a more meaningful teaching of Mathematics, the Queen and Servant of the Sciences" (SMSG, 1966a, n.p.). In the new curriculum reform efforts mathematics' "service" to the sciences was embedded in a new way of thinking about math and science as a norm of rationality deemed useful in organizing social projects for planning America's postwar society. As the previous chapter discussed, mathematics was given new cultural value as a scientific way to measure, calculate, and model a society envisioned as inclusive, equitable, and democratic.

Whereas the emphasis on math taught at all levels signified a newly emerging focus on the elementary math curriculum, it was not simply about young children learning math. Elementary school mathematics—and math education in general—were taken as integral to living as "intelligent citizens" (National Council of Teachers of Mathematics [NCTM], 1945). Within the historical context of reshaping society, the "intelligent" kind of citizen appeared as a kind of person who could be fabricated through the remaking of mathematics as something new.

The new curriculum reforms circulated within a new way of thinking about school science and mathematics as tools for making a modern, intelligent, and rational kind of citizen. Within this logic, a different relationship emerged between planning for social change and the school curriculum. This was evident in the reorganization of curriculum planning and development projects as professional collaborations among scientists, mathematicians, educators, and psychologists involved in curriculum design, textbook writing, and instructional planning.

The curriculum texts by the SMSG quoted earlier and examined in the next sections signify one such collaboration. Historically, this group played a significant role in the creation and circulation of what the new math curriculum movement represented by and large.[1] Given the belief that math education mattered to the development of the nation and its future, the curriculum reforms that were funded, created, and circulated tell a story about the kind of nation the United States thought it should be after World War II and how one ought to live as an intelligent, rational, and democratic citizen.

These presumed rational modes of thinking and acting were associated with a so-called democratic kind of person and defined through postwar social sciences and cognitive psychology. Through the use of models to pattern and predict what were thought of as universal human thought processes, behaviors, and attitudes, cognitive psychology defined who the democratic person was and should be. Elaborated in the previous chapter, the democratic citizen was given the traits of an open mind, tolerance, imagination, autonomy, creativity, and flexibility (Cohen-Cole, 2009). More than psychological traits, these cultural ideals were ascribed to individuals through what were seen as civic behaviors and democratic attitudes. In the sections that follow, the chapter illustrates how these cultural norms were embedded in the new elementary math curriculum as rules for how children should think and act as part of *math for all.*

Reading the curriculum will highlight how the new elementary mathematics was not merely a copy of what modern mathematicians were doing. Mathematics was translated into math education via the principles of cognitive psychology so that it would seem to make sense for all children to internalize the patterns of thinking and acting associated with rationality, intelligence, and democracy. In the translation, children were seen as having common psychological traits that presumed them all as capable of mastering the structure of mathematics. These traits, given in the language of a psychology that orders who children are and how they learn, are read as part of the process of "alchemy" (Popkewitz, 2004) or translation. It is through the translation that mathematics education could become a tool for making the modern life by making the child into a certain kind of citizen through mathematics. The SMSG curriculum texts are examined as expressions of the modes of reasoning that related children's learning to notions of social progress and expressions of citizenship.

The analysis of the math problems in children's textbooks intersects with such psychological "truths" about how children learn. Important texts, such as *A Study of Thinking* (Bruner, Goodnow, & Austin, 1956), *The Process of Education*[2] (Bruner, 1960) and *The Psychology of Intelligence* (Piaget, 1950) are explored in terms of how the logic of cognitive psychology operated to give equivalence to *all children* to rationalize how they were to learn and understand the content of the new curriculum that was more than just mathematical thinking. These psychological works and the norms they carried were imported into the American curriculum to articulate a new kind of child for a hoped-for future society. In this process, the cultural principles of creativity, flexibility, and independence operated through cognitive psychology to identify the rules by which to make modern mathematics intelligible to children and fabricate them a part of a democratic and inclusive *all*.

By making visible how the logic of equivalence and equality embedded in the equal sign in the new curriculum overlapped with psychological modes of identifying children and characteristics of their thinking, the next sections explore how a new mathematics was organized to include *all students* in modern math and life. Attention is given to the "mathematical competences" that embodied a way of reasoning about equivalence and inclusion that were, on one hand, to help "all students, with their varying interests and abilities" understand the underlying structures of mathematics (Begle, 1958, p. 616). On another, they embodied particular civic virtues "required for effective and intelligent living in our culture" (CEEB, 1959, p. 7) that distinguished who could be part of *all students* as citizens living the modern and intelligent life.

Set Theories of the New Curriculum

The introduction to a first-grade math text published by the SMSG stated that "one of the prerequisites for the improvement of teaching mathematics in our schools is an improved curriculum . . . which reflects recent advances in mathematics itself" (SMSG, 1966a, n.p.). Set theory represented one of these advancements and was to provide children with an understanding of the structures and patterns that organized numerical relations and operations. Its placement in the curriculum was "an effort to help children in learning to appreciate the precision in making statements to describe things" (SMSG, 1961, p. 1). This appreciation was linked to helping children understand that "the language of sets is an effective way to express ideas, not only in mathematics, but also in many other fields of knowledge" (ibid).

The specificity of language use expressed the belief in and importance given to mathematics as an accurate and effective way to see and communicate about the world. The use of the language and its practices embodied a particular mode of living as an intelligent kind of person. This will be seen in how the emphasis given to associating precise vocabulary with mathematical concepts of equivalence embodied a way of establishing equivalencies among children as creative, expressive, and inventive kinds of people through the language of cognitive psychology.

The precise use of vocabulary was emphasized throughout the first-year student and teacher texts—particularly as a way to understand the mathematical notion of <u>equivalence</u>.[3] <u>Equivalence</u> and the exact terminology that gave it meaning were underscored as a way to prepare children for the notion of number and the inequality of number (SMSG, 1963, p. 4). In the first years of mathematics education, children were to make sense of the comparative logic of equivalence expressed in the language of <u>more than</u> and <u>less than</u>. Before interacting with numbers and symbols, like the equal sign, children were expected to use specific language to compare and establish equivalence and nonequivalence between <u>sets</u> of objects. In the student text these relationships were to be determined through a description and comparison of visual representations (trees, birds, and kites) in response to the question: <u>Are the sets equivalent?</u>

According to the comparative rules embodied in the language, it seemed that the sets are <u>not equivalent</u> because there were <u>more members</u> in the <u>set</u> containing trees. This expression was more than descriptive. Embedded in the use of language to understand equivalence is a particular way of thinking that can be characterized as grouping objects into classes (sets) and responding to them (describing them) in terms of what they have in common. Making things equivalent expressed a comparative logic that could be taught to children.

Yet through the use of psychology as a tool to think about the qualities and characteristics of mathematical knowledge, the making of equivalence also seemed to entail a mode of thought that involved an "act of invention" (Bruner, Goodnow, & Austin, 1956, p. 2). This act of invention that appeared to be involved in grouping things together to determine equivalence was related to the "imagination" required for developing the concept of number (SMSG, 1966b, p. 1). By identifying and making the four trees into members of a set, children should also "imagine" many other sets of objects signifying that number. This act of invention was not merely a free association of things into patterns but had certain rules for ordering and classifying a relationship of sameness or difference.

So inventing equivalent groups as sets of objects was not only to do with mathematical reasoning. This is clear in how the link between language use and imagination was bound to "evidence for the belief that a student who understands the solutions to systems . . . can create his own methods" (UICSM, 1956 quoted in Bidwell & Clason, p. 658). It seems that if students could systematically see and describe relationships among objects, they would be able to think creatively and independently about those objects.

Whereas teachers were supposed to "encourage the child to use his own method of thinking," this thinking was not independent from the modes of thought embodied in the manipulation of language, images, and physical objects for understanding equivalence (SMSG, 1966c, p. 400). The creation of "his own method" was organized through the ordering principles of equivalence for seeing objects in the world as part of the same set. Imagination, creativity, and independent thinking appeared as mathematical skills that worked toward building mathematical competences.

This can be seen in a new mode of representing relationships of equivalence through the use of the equal sign that embodied a supposedly imaginative method for answering "How many?" on one's own. Prompted by the directive, "Write the equation," children were supposed to—in the language of the text—"imagine" the *set* of objects as *three* and *two* and *join* them together. Creating a solution to the problem of equality expressed by ____ + ____ = ____ required a way of reasoning about distinct objects (a balloon, tree, ice cream cone, a bike, and a monkey) as members of a set and imagining an identity that would create an equivalence between them.

This way of thinking about equality was about more than mathematical competences in organizing and understanding equivalences. The rules for grouping objects together were bound to the rules organizing how to see and talk about the equivalence and nonequivalence between objects in the world. This math skill was understood through the psychological inscription of all children's imaginative capabilities to see the objects and assign a relationship between them. The problem of creating a symbolic expression of equality between 3 + 2 and 5 also worked to establish equivalences among children in that they were given traits as creative, imaginative, and precise in their use of language and methods for solving new problems in school mathematics.

A cultural ideal of technically producing creativity and imagination through the use of mathematics as a tool for making a democratic nation was embedded in the reorganization of the math curriculum. The pedagogical practices embodied in the new mathematics textbooks related particular mathematical competences to cultural standards required for living as an intelligent citizen, capable of seeing and creating solutions to problems in the world. This ordering of conduct had less to do with the mathematics of equivalence and more with the inscription of ways of reasoning about equivalences between children's thought processes and expressions of rationality and intelligence. These cultural norms of creativity and imagination were translated through the principles of psychology to inscribe the "right" conditions for living and learning as intelligent and rational citizens.

Materials and the Materiality of Abstract Thought: Solving for the Unknown

Further into the first-year text, children would encounter open addition and subtraction expressions as the core of their problem-solving. These problems were presumed to be more abstract with that abstraction expressed through the symbolic logic $3 + 0 = \square$ and $2 + 2 = \square$. Acknowledging that children might not be able to think abstractly, physical objects were presumed useful to model their problem-solving. It was thought that "working with concrete materials is essential as it provides every child with a method for finding the sum" (SMSG, 1966a, p. 214). In other words, these objects, like blocks, pennies, or toys, embodied a strategy for finding the unknown if used correctly (ibid, p. 236). "Regardless of intellectual level," children could join together sets of objects to systematically solve the problems (SMSG, 1966b, p. 2).

This method was not merely a way of working with the curriculum materials to produce understanding. It also embodied a cultural expression of the kind of child who was capable of thinking abstractly and the mode of thought that made that possible that had to do with the psychology of the child rather than the mathematical knowledge itself. In reorganizing the curriculum, it was assumed that "what is most important in teaching basic concepts is that the child be helped to pass progressively from concrete thinking to the utilization of more conceptually adequate modes of thought" (Bruner, 1960, p. 38). With "concrete thinking" as one of the limit points of the trajectory of a "natural thought process," a more abstract form of thought was presumably the other limit. Abstract thinking was given as a higher and more adequate form of thinking and observable as a kind of "mental mobility" (ibid, p. 42). This mental mobility required a child to see beyond the immediacy of the concrete materials while thinking with them.

The "considerable and extensive manipulation" of objects was to serve as a bridge to a presumably more adequate mode of thought that no longer relied upon an interaction with the physical environment. The identification of objects with images and numbers was thought to move children from a concrete world tied to the specificity of things to a more general and abstract mode of reasoning. Solving problems like $2 + 2 = \square$ entailed the use of materials for children to model what is known to find out what is not. This mode of thought was characterized in cognitive psychology as "a way of arranging knowns and unknowns in equations so that the unknowns are made knowable" (Bruner, Goodnow, & Austin, 1956, p. 7). The unknown could now be seen.

This way of thinking about knowns and unknowns related the manipulation of set objects in the curriculum to a cultural way of reasoning about the systematic planning of the future. In *Excellence: Can We Be Equal and Excellent Too?* (Gardner, 1961), a concern was that

we don't even know what skills may be needed in the years ahead. That is why we must train our ablest young men and women in the fundamental fields of knowledge, and equip them to understand and cope with change. That is why we must give them the critical qualities of mind and the durable qualities of character that will serve them in circumstances we cannot even predict.

(ibid, p. 53)

The presumed qualities of mind and character were part of the curriculum as traits given to children seen as open to the unknown and methodical in their problem-solving. Inscribed in the use of materials as a method of problem-solving, these traits that children were to internalize were assumed to be useful in solving future and unforeseeable problems. Variations on the same set of rules, the materials were to aid children in methodically modeling what could not be known with things that apparently could.

This independent and abstract thought was presumed to be in development, just as creativity was. Using the objects, then, was to take thought beyond the immediacy of objects, where there was "more than reality involved, since the world of the possible becomes available for construction and since thought becomes free from the world. Mathematical creativity is an illustration of this new power" (Piaget, 1950, p. 151). Pedagogically, this "mathematical creativity" was to incite thinking about equality and equivalence that was not tied to the immediate and concrete world and was given meaning through the psychological traits ascribed to creativity and abstract thought, free from the material objects. Presumed as a developmental process of the mind, teaching rested on the assumption that each child could be taken from a way of thinking about the world tied to objects to a mode of thinking freed from them—as an expression of a mathematical form of creativity.

Thinking independently, finding their "own" methods, and freeing thought do not merely exist as a universal logic. These characteristics of thought came to make sense within the cultural and historical context wherein the qualities ascribed to democratic and intelligent citizenship were valued as traits for children to internalize. The independence, creativity, and freedom were not given in that children could use the materials however they wanted. Like the use of language and acts of invention involved in understanding equivalence, the materials involved in solving expressions with the equal sign and numeric symbols entailed some rules.

These rules were mathematical in terms of organizing modes of thinking to solve inequalities. But they were also cultural with regard to how they ascribed identities to children. That is, using the materials seemed to require a "thoughtful" and "meaningful" manipulation on the part of the child (SMSG, 1966b, p. 2). This use could be understood in relation to the rule that "with the meaningful manipulation of concrete objects, children become active participants and not merely passive watchers or listeners" (ibid). So, through the use of materials, children were to be seen as more active in their learning—as a method of moving from concrete to presumably more "mobile" forms of thinking and finding unknowns from knowns.

In this reorganization of the curriculum around notions of an abstract and mathematically creative mode of thought, the equal sign embodied notions of equivalence and equality that the child was to understand through the use of materials. Their use was organized according to a logic that operated through cognitive psychology to classify certain kinds of thinking as higher, more adequate, and aligned with the creativity that was granted cultural importance. Interestingly, the classifications of thought that were embedded in the use of the equal sign gave expression to children's unknown and internal thought processes—as knowable, flexible, abstract, and independent.

Reversibility of Thought and Modeling Abstraction

At the end of the first-year SMSG texts, children were held to a new standard of thinking implied in a different kind of problem. The child was to undo, or mentally reverse, addition and subtraction by relating problems like $8 + 1 = \square$ and $9 - 1 = \square$.

No longer relying on objects and images, the reversal relied upon what were given as "more conceptually adequate modes of thought" (Bruner, 1960, p. 38) as the expression of a higher form of reasoning. This reversal of thinking expressed through problems like those here represented a presumed psychological trajectory toward what was given as the outer limit point of human thought on a continuum from concrete to the abstract. In that, the problems resembled and were modeled by what were taken as the highest levels of the thinking process—abstraction.

Granted as a psychological truth, abstract thinking was given a "scientific" meaning in reference to the "reversibility of thought" (Piaget, 1950, p. 42). The undoing of mental processes, or reversing "intellectual operations," was believed to represent the highest form of thinking as an observable thing through solving math problems like those presented here. Undoing the order of things to solve for the unknown in both problems was thought to move children from a mode of thinking wherein changes, like adding 1 to 8, are flexible and reversible. The logic embodied in the use of the equal sign organized a rule of reasoning whereby children's thinking seemed to move toward a higher form of thought that was to be seen as abstract and mathematically creative in its reversibility.

The importance given to this ability to undo or reverse thought in the new curriculum was understood through the language and tools of cognitive psychology in making a notion of abstract thought visible. In modeling and defining abstract thought—as a higher form of thinking equated with a mode of "mathematical creativity"—mathematics was also in service to cognitive psychology. That is, mathematics provided a symbolic logic through which to represent what was given as a reversible thought process: $y - x = x'$ or $y - x' = x$ (Piaget, 1950, p. 42).

This mathematical model of reversibility of thought worked as a rule of cognitive psychology to establish equivalences among children, grouped together by presumably common and universal models of thinking. As a fundamental part of the internal thinking "mechanism," this process was presumed to not yet function at this highest, most perfect form in young children (Piaget, 1950). Nevertheless, children appeared to have the same internal and innate structure underlying the development of a thinking process that could follow a predictable pattern from concrete to abstract.

The psychological and mathematical expressions of reversibility of thought inscribed a generalized pattern of the development of abstract and logical reasoning that seemed to repeat over time and in all people. In making apparent an otherwise unseen thought process, the equation provided rules for how to organize the curriculum to bring the underlying structures of mathematics "within the intellectual reach of children of all ages" (Bruner, 1960, p. 32).

With this equivalence established, it seemed reasonable to think that *all children* could master the foundations of a new and modern mathematics to function according to the same operations and ways of reasoning. In mapping this pattern of thought, the principles of mathematical and psychological equivalences organized the curriculum and children's thinking on a continuum from seemingly lower to higher forms. Following a linear sequence of mathematical practices and rules in the curriculum, all children's thinking could presumably develop toward reversibility, abstraction, and creativity as expressions of rationality and intelligence.

The problems like those in this section appeared to model an internal thought process and general pattern of a reversible style of reasoning that could be developed in children. Yet patterns of reasoning about the equal sign in the new math curriculum did not only inscribe rules for understanding mathematical notions of equivalence and equality. They also carried cultural principles about the identity and equivalences of children as creative, abstract, flexible, and independent. More than mathematical thinking, these were traits that would identify children as intelligent citizens of the nation.

Whereas the elementary math curriculum ordered the competences and traits identifying children as an equal part of the *all*, it also made visible the differences by which to divide them. The next section explores how cultural theses of in/equality emerged as new ways of classifying the children who did not seem to belong to the all.

New Math and the Cultural Construction of In/Equality

> If all students are helped to the full utilization of their intellectual powers, we will have a better chance of surviving as a democracy in an age of enormous technological and social complexity.
>
> (Bruner, 1960, p. 10)

(Re)visioning and "surviving as a democracy" were apparently tied up in all students' development for the future. This development was not only about technological or economic innovations. It was also tethered to new ways of thinking about the place of equality and opportunity in the reorganization of schooling. Here the impulse to survive through a new mathematics also embodied social and cultural concerns with planning a more egalitarian society in which all citizens had the opportunity to learn and live the modern life, as discussed in the beginning of the chapter.

In the postwar United States, educational opportunity was increasingly considered integral to assuring all Americans the chance to a better life. *Brown vs. the Board of Education* invoked the public conception of the importance of equal educational opportunity as well as education as the right of a free citizenry. This consensus on the role of education in producing social equality related to the belief in public welfare as a social policy concern. These ideals were expressed through federal policies aimed at reorganizing education with opportunity and equality as a central concern. Situated within the War on Poverty and the Great Society social reforms in the United States, the Elementary and Secondary Education Act (ESEA) was formulated as a public law in 1965. With a primary focus of granting educational rights, access, and opportunity, the concern was in "meeting the special educational needs of educationally deprived children" (ESEA, 1965, p. 27).

The "educationally deprived" appeared to express a new kind of social problem in poverty as a cause of educational inequalities. Ordered by a comparative way of reasoning about the right to an equal and quality education for all students, the law embodied cultural theses about *who was not equal*. The representation of self and other, from which inclusion means access to opportunities, also expresses a continuum of values that differentiates. To the extent that this problem could be solved and the "educationally deprived" could become part of the all, this child would need to be identified.

This section examines how the new categorization for *all* served to classify the children who were included as part of the all while simultaneously distinguishing the nonequivalent group for which the right to an equal and quality education was named. It will elaborate how equality and equal educational opportunities were to be granted on the condition of all children internalizing the cultural norms of creative, reversible, abstract, and independent thought processes. These kinds of children, perceived as citizens who participated in school and society according to these rules, would already have the right and opportunity to educational e/quality outlined in the policies and curriculum.

On the surface, the policy solution to educational deprivation rested in creating the "right" learning environments by reallocating funding, resources, and materials to reduce inequality in terms of economic access and opportunity. As a cultural distinction, not just an economic one, the inequalities associated with "culturally" and "educationally deprived children" did not simply rest in the environment but were *seen as problems due to differences in the child*. Educational success for "these" children seemed to require a reversal of "the course of intellectual retardation in the culturally deprived" (Ausubel, 1963, p. 464). This concern was made explicit in the introduction to a revised edition of an SMSG first-year text wherein "culturally deprived children" emerged as a distinct group with whom the curriculum was concerned (SMSG, 1966a, preface). The preface of this revision articulated the hope that the new edition would better meet the needs of the "disadvantaged" and "deprived." Embedded in this hope was a fear of children who seemed to lack the mathematical competences to participate in school and society.

The logic of the policies and interventions aimed at reforming the group distinguished as "culturally and educationally deprived" embodied a goal of improving the unequal conditions of schooling. Yet the deprived and disadvantaged distinctions emerged in relation to the child who embodied the traits that codified the intelligent and rational citizen. This "disadvantage" was expressed as psychological traits, perceived as different forms of rationality, intelligence, and participation.

An article titled "The Overlooked Positives of Disadvantaged Groups" articulated the "disadvantaged" as a group understood according to the "mental style or way of thinking characteristics of these people" (Riessman, 1964, p. 225). Given psychological distinctions, this group could presumably be understood in reference to several defining qualities. One distinction granted that "slowness is an important feature of their mental style" (ibid). The slowness with which deprived children apparently performed intellectual tasks could be understood in contrast to the child whose "mental style" and thought process was precise and methodical. Slowness was attributed to this child representing a "physical style of learning" and a thought process that "persists in one line and is not flexible or broad" (ibid, p. 226). Both of these characteristics of deprived children are given in reference to children who think beyond the physical realm, in the abstract, as well as those who are able to be reverse their thinking.

Whereas the physical style was given attention in the curriculum through the use of objects and materials in solving for equality and understanding equivalence, the presumption was that this was a method to progress a natural trajectory of thought toward abstraction—a mode of reasoning the deprived kind of child seemed to lack. Distinguished from abstract, flexible, and efficient styles of thought inscribed as the standard in the curriculum, one-track thinking tied to the physical world would be one way to identify this group of children as impoverished and part of the problem of educational inequalities.

Interestingly, the "single-minded" quality was articulated in reference to notions of creativity. Like flexibility of thought, open-mindedness, and abstract thinking, one-track thinking seemed to offer a considerable potential for creativity (ibid). This notion of creativity was inscribed as a different kind for the child seen as disadvantaged. "The mental style of the socially and economically disadvantaged learners resembles the mental style of one type of highly creative persons. Our schools should provide for the development of these unique, untapped national sources of creativity," (ibid, p. 231). This "type" of learner gives a name to distinguish the presumably different, nonequivalent group for which the right to an equal and quality education was named.

This apparently different form of creativity was articulated in relation to other features identifying the thought processes of those considered deprived. In "The Culturally Deprived Child" distinct creative modes of thinking were characterized as a problem-centered rather than abstract-centered approach, the freedom from being word-bound, an equalitarianism in informality, the avoidance of competitiveness, an emphasis on collaboration over individualism, a spatial rather than temporal perspective, and the use of a physical and visual style in learning (Riessman, 1962). This kind of child was identified as not being articulate or precise with words as well as informal, dependent, and concrete in thought. These traits and the different kind of creativity they presumed were framed as a deficit compared to the other, presumably intelligent modes of creative thinking. That is to say, the child who was not correctly and independently using the physical materials, following the rules bound to language use, formally seeing equivalence and equality in the images and set objects, developing an abstract and higher order of thinking, or using efficient problem-solving methods would not develop the mathematical modes of creativity ascribed to an intelligent citizen.

The line of demarcation between the deprived and intelligent was drawn along psychological, economic, geographical, and racial lines. This can be seen clearly in an article titled "Who Are the Socially Disadvantaged?" (Havighurst, 1964). In a special issue of the *Journal of Negro Education* about educational planning for socially disadvantaged children and youth, socially disadvantaged was synonymous with the culturally and intellectually deprived and indicated a general disadvantage for living in society. The child classified as belonging to this impoverished, lower-class culture was presumed to develop a "different kind of mind" (ibid, p. 211). This difference was expressed as "certain personal deficits" and were made visible as a restricted use of language, inferior auditory discrimination, inferior visual discrimination, inferior judgment concerning time, number, and other basic concepts" (ibid, p. 214). This child had apparently not developed the ability to listen, see, pay attention, or talk about the world in particular ways.

These perceived psychological deficits were primarily thought to be a factor of the environment in terms of the conditions in which children grew up. Careful not to identify economic factors and poverty as the only cause of deprivation among "working-class children," the differences were also attributed to geographical and social distinctions. Given that "there is substantial doubt that the socially disadvantaged children in our big cities have any positive qualities of potential value in urban society," the children who were seen at the largest disadvantage to participate fully in the democratic society were those living in large cities (ibid, p. 215). It seemed that "the socially disadvantaged children tend to come from families that are poor and recent immigrants to the big cities" (ibid).

The geographical distinction of "urban" was also a cultural way of characterizing a certain kind of people who lived in the big city. Drawn along racial and ethnic lines, the disadvantaged would be characterized as Negros and Whites from the rural south and Puerto Ricans, Mexicans, and European immigrants from rural areas (ibid). Perceived as lacking the tools to live as intelligent citizens, the deprived were distinguished by a perceived disadvantage for full participation as rational, creative, independent, and open-minded citizens.

The categories that defined one's opportunities to be an intelligent citizen through *math for all* also defined who was not that child. The differences were established in terms of how the child reasoned, communicated about, and interacted with the world through mathematics. Whereas the differences were seen as a problem in children to be addressed in the curriculum, they were paradoxically produced within moves to achieve more equitable social opportunities for *all*.

Progress, Prosperity, and Equality = Children's Mathematics

Presently it seems so obvious that children's mathematical reasoning is linked to their participation in society. Take, for example, the contemporary discussions about "mathematical literacy" in the Organization for Economic Cooperation and Development (OECD, 2009) wherein "mathematical literacy is an individual's capacity to identify and understand the role that mathematics plays in the world, to make well-founded judgments, and to engage mathematics in ways that meet the needs of that individual's current and future life as a constructive, concerned and reflective citizen" (p. 14). Learning mathematics is not only about math. It embodies rules for living as the kind of citizen that uses math as a mode of organizing a constructive, well-founded, and engaging life.

Drawing attention to how the elementary math curriculum was about much more than mathematical competencies, this chapter has explored how its reorganization worked to make children carriers of the modern life— inscribed in the rules of creativity, independence, and flexible thinking as cultural rules by which to live. Emerging as the image of rationality and progress, children were to internalize the psychological traits as norms of citizenship and rules for belonging as part of *all*. In determining the standards of inclusion into the *all*, the very rules of equality and equivalence that organized who was the intelligent citizen also classified who was not that— distinguishing economic, social, geographical, and racial characteristics of "deprived" and "disadvantaged" children who seemed to embody different forms of living as the expression of inequality.

The investment in children's math education continued to assemble with social commitments to provide equal opportunities for *all*. New descriptions and calculations of student under/achievement, in/equality, and in/competence would emerge. Within the context of these concerns, the new math curriculum would become the target of pointed criticism in the early 1970s. In the period of reform that followed, failure in mathematics was not historically new. What was new were the ways in which it would be articulated as a problem to solve at the individual level as the differences in children's abilities and motivation in learning math. As a historical shift from a new mathematics, the reforms that came next would seem to disarticulate math education from notions of citizenship. Yet they very much relied upon the historical associations that tied elementary math education to notions of national progress through the development of the child as a rational and intelligent kind of citizen.

Notes

1. The texts and problems analyzed are not meant to be representative of a singular new math movement. As argued in *New Math: A Political History* (Phillips, 2015), one approach to *the* new math did not exist. There were varying perspectives as to what math was for, who should learn it, and how. Yet there was a consensus formed around the belief that math education mattered to the development of the nation and its people.
2. This text was written as the result of a 1959 conference at Wood's Hollow among "experts" in the fields of education, learning sciences, science, and mathematics. The National Academy of Sciences, the RAND Corporation, and the U.S. Air Force funded the collaboration aimed at creating curriculum and improving the teaching of math and science.
3. The underlined text is representative of the vocabulary that appeared in the curriculum.

References

Ausubel, D.P. (1963). A teaching strategy for culturally deprived pupils: Cognitive and motivational considerations. *The School Reviewer*, 71(4), 454–463.

Begle, E.J. (1958). The school mathematics study group. *The Mathematics Teacher*, 51(8), 616–618.

Bruner, J. (1960). *The process of education*. Cambridge, MA: Harvard University Press.

Bruner, J., Goodnow, J.J., & Austin, G.A. (1956). *A study of thinking*. New York, NY: John Wiley & Sons, Inc.

Cohen-Cole, J. (2009). The creative American: War salons, social science, and the cure for modern society. *Isis*, 100, 219–262.

College Entrance Examination Board. (1959). Program for college preparatory mathematics. In Bidwell, J.K. & Clason, R.G. (Eds.) (1970), *Readings in the history of mathematics education*. Washington, DC: National Council of Teachers of Mathematics. pp. 664–706.

Dow, P.B. (1999). *Schoolhouse politics: Lessons from the Sputnik era*. Cambridge, MA: Harvard University Press.

Elementary and Secondary Education Act. (1965). P.L. 89–10. *United States Statutes at Large*, 79, 27–58.

Gardner, J.W. (1961). *Excellence: Can we be equal and excellent too?* New York, NY: W.W. Norton & Co., Inc.

Hacking, I. (2002). Inaugural lecture: Chair of philosophy and history of scientific concepts at the College de France. *Economy and Society, 31*(1), 1–14.

Havighurst, R. (1964). Who are the socially disadvantaged? *Journal of Negro Education, 33*(3), 210–217.

National Council of Teachers of Mathematics. (1945). The second report of the commission on post-war plans: The improvement of mathematics in grades 1–14. *The Mathematics Teacher, 38*, 195–221.

Organization of Economic Cooperation and Development (OECD). (2009). *PISA 2009 assessment framework: Key competencies in reading, mathematics, and science.* Washington, DC: Author.

Phillips, C.J. (2015). *The new math: A political history.* Chicago, IL: University of Chicago Press.

Piaget, J. (1950). *The psychology of intelligence.* New York, NY: Harcourt, Brace & Co., Inc.

Popkewitz, T.S. (2004). The alchemy of the mathematics curriculum: Inscriptions and the fabrication of the child. *American Educational Journal, 41*(4), 3–34.

Popkewitz, T.S., Diaz, J., & Kirchgasler, C. (2017). *A political sociology of educational knowledge: Studies of exclusion and difference.* New York, NY: Routledge.

Riessman, F. (1962). *The culturally deprived child.* New York, NY: Harper & Row.

Riessman, F. (1964). The overlooked positives of disadvantaged groups. *The Journal of Negro Education, 33*(3), 225–231.

School Mathematics Study Group. (1961). Mathematics for the elementary school: Concept of sets. Teachers' commentary.

School Mathematics Study Group. (1963). Mathematics for the elementary school: Book 1. Teachers' commentary (Preliminary Edition).

School Mathematics Study Group. (1966a). Mathematics for the elementary school: Book 1, Part 1. Teacher's commentary.

School Mathematics Study Group. (1966b). Mathematics for the elementary school: Book K. Teacher's commentary (Revised).

School Mathematics Study Group. (1966c). Mathematics for the elementary school: Book 1, Part 2. Teacher's commentary (Revised).

University of Illinois Committee on School Mathematics (UICSM). (1956). The University of Illinois school mathematics program report. In Bidwell, J.K. & Clason, R.G., (Eds.) (1970), *Readings in the history of mathematics education.* Washington, DC: National Council of Teachers of Mathematics. pp. 655–663.

5 Civil Rights, Fears of Failure, and the Motivation of Basic Skills

Fabricating the Mathematically Dis/Abled Individual

With the emergence of *math for all* reforms in the middle of the 20th century came a new common sense: Learning math is important for making the child into an intelligent and rational citizen of the Unites States. Through learning mathematics, children were given ways to see, think, talk about, and behave in the world as flexible, creative, and independent thinkers who could precisely use the language of math and its underlying structures to solve unknown problems. Clearly, mathematics in schools was about more than mathematical reasoning.

In contemporary reforms, children's math learning continues to be tied to their role as citizens and is presently expressed in terms of how math education in the 21st century "will serve us well in becoming a society where all citizens are confident they can do math" (National Council of Teachers of Mathematics [NCTM], 2000). In this continued promise of *math for all* students—in service to the nation—is a historical way of thinking about the relationship between children's math education and their identities as certain kinds of people. In this continuity, however, there are changes as further nuance was given to determine who could be part of *math for all*.

This chapter continues to attend to how the logic of equality embedded in the use of the equal sign in the elementary math curriculum intersects with and orders ways of thinking about how to (re)organize *math for all*. It focuses on how the equivalences given to make children as part of the all were defined and produced new ways to think about children during the math reforms in the United States during the 1970s and 1980s, referred to as a "back-to-basics" movement. Generally speaking, in the history of math education, the back-to-basics movement does not get much attention. Yet it served as an important event in the history of working to organize math for all children.

During the call for a move back to basics, the common sense that all children needed to and could learn math continued to be visible in elementary math education reforms. In the October 1969 issue of the *Arithmetic Teacher*, a new footer appeared along the bottom of each page that read, "Excellence in Mathematics Education—For All." As if signaling the ways in which mathematics education had once again become an issue of educational reform, "excellence—for all" seemed to demand something new.

As a shift from the previous new math reforms, the back-to-basics movement inscribed new identities for children and their mathematics—changing from rules of intelligent citizenship given as modes of creativity, independent thought, and flexibility to traits of individuality defined through notions of personal interests and motivation. The moves toward a basic math education provided new ways of classifying children as individuals with identities and differences expressed as their own abilities for participating in school and society. The focus on the fabrication of the child as an individual highlights how the norms of participation embodied in the math curriculum were still related to previously inscribed notions of citizenship but given different classifications.

Emphasis on the individual and the differences that defined one's individuality was not only visible within math education. In other social contexts, the individual was given attention as a basic and fundamental component of social progress within a web of economic, political, and cultural challenges about rights, equality, and freedom. Underlying what constituted these challenges was a "spirit of egalitarianism and inclusiveness that rejected traditional hierarchies and lines of authority, asserting instead the equality of all people, particularly women, gays and lesbians, people of color, and the disabled—that is, the majority of people" (Borstelmann, 2011, p. 3). Here the representation of individual identities served as a way of thinking about in/equalities through distinctions of difference.

Individuality became integral to expressing notions of difference as embedded in thinking about inequality, egalitarianism, and inclusion. The differences and inequalities that shaped individual's social, political, and economic challenges were simultaneously troubled and presumed to be mutable. Unequal results and differential opportunities were taken to be the result of, and therefore altered by, individual actions. Importantly, this interest in the individual also encoded new norms for discovering, identifying, working on, and planning the self (Bellah et al., 1985; Turner, 1976; Zurcher, 1977). Working on the "self" was to be in the name of gaining a greater sense of freedom, finding one's place in society, and inspiring one's own commitments to personal fulfillment.

This chapter will highlight how cognitive and social psychologies provided classifications as governing practices in making the child and working on the self that become visible in the curriculum as the tools of translation to reorganize the image of the child. It will continue to elaborate how math education operated in the making of people and how the cultural thesis about the child who could be part of *excellence for all* produced differences that have little to do with math. In relating a seemingly basic understanding of the equal sign in the curriculum to new ways of reasoning about identities and differences of each child, the analysis centers on how math in schools was (re)arranged and placed particular skills, abilities, and modes of thought as "basic" and "natural" traits of the child as a self-regulating, self-motivating, and self-sufficient kind of individual. These traits of individuality and sense of self were inscribed in relation to the *all* yet give the child his [sic][1] own identity as well.

Attending to the Basics in Math: The Individual, Intuition, and Interest

Although the back-to-basics movement could be considered a shift away from the previous new math curriculum, it also expressed important continuities in terms of a fundamental link between math education and the reorganization of social and cultural life. What distinguished the back-to-basics movement, however, were new modes of living this life.

No longer simply about making intelligent citizens, mathematics education was related to "the development of life (or survival) skills—that is, competencies needed for personal growth and for successful existence as a citizen, consumer, jobholder, taxpayer, and member of a family" (Brodinsky, 1977, p. 524). Implicit in this statement of math as integral to life skills is a double layer of competencies whereby knowledge learned in school was related to one's participation in society as an individual—given multiple forms of expression and responsibility that moved beyond citizenship. This shift toward math's basic skills as modes of personal survival and growth shaped what basics to get back to.

Situated in a growing discomfort with previous reforms, a prominent text, *Why Johnny Can't Add: The Failure of New Math* (Kline, 1973), suggested that the emphasis on structure and abstraction was the reason, not only for "Johnny's" failure but also for math's inadequate contribution to the development of science and society. New math had apparently failed not only society but children as well. This new focus on children, named in the mathematics and psychology texts as individuals (Johnny, Kenneth, Bob, Caesar, and Leslie), marked an emerging impulse to situate mathematics education in each child. This individuality expressed in the names became part of the construction of the kind of person that was to be produced through math in schools. More than a name, it signifies how the psychological rules were to be internalized in each child and given expression as a mode of living and surviving.

With a shifting focus on individuality and attention on the presumed failures of new math, there emerged a new common sense of what would constitute the basic education for life in a progressive and democratic society. A basic education, as the presumed foundation of success in math for all children, was to move away from abstraction toward a more "interesting" and "useful" mathematics that did not "destroy the spirit and life" of mathematics (ibid). A more "natural motivation" appeared to be found in problems that would "revivify math by the air of reality" (ibid, p. 149). Math in its most basic form was to be lively, motivating, and interesting—not abstract or removed from what apparently constituted the problems of real and everyday life.

In this association of math with "natural" interests and motivations, it seemed evident that "the basic approach to all new subject matter at all levels should be intuitive" (ibid, p. 157). A basic mathematics curriculum was to be built upon what were taken as inherent modes of reasoning, instinct, and intuition. This supposedly natural intuition was thought to provide the opportunities for success with the basic curriculum and life. Expressed in research concluding that even "infants have ample opportunity to learn about number, repetition, regularity, differences in magnitude, equivalence, causality, and correlation," the notion of a mathematical intuition established an equivalence in children through which they were all seen as capable of the same basic mathematical insights and skills (Ginsburg, 1977, p. 30).

The assumptions of a natural intuition and motivation that organized the basic math curriculum to provide equal opportunities for excellence were not biological distinctions. Rather, they were cultural ones. This distinction between nature and culture is expressed as unproblematic in the following discussion of the math curriculum:

> it is hard to see how any child, rich or poor, Western or non-Western can grow up in an environment which does not offer him or her the natural opportunity to learn about basic aspects of quantity. In this sense, environments are all similar and quantity is universally available. We shall see later that there is a sense in which all children take advantage of this natural opportunity.
>
> (ibid, p. 31)

Articulated as skills and opportunities that would develop in any environment, this notion of "natural opportunity" is a cultural expectation wherein the idea of a basic, intuitive mathematics was apparently available to all children.

At the same time that it was possible to think that universally all children could naturally excel in a basic form of math, it appeared that some children did not adequately "take advantage" of the "natural opportunities" to be a part of school mathematics. When Kenneth, a child interviewed to better understand his "challenges" with math, "interpreted the + and = in terms of actions to be performed," his interpretation was seen as one that "can lead to trouble":

I: How would you read this [$\square = 3 + 4$]
K: ... Blank equals 3 plus 4.
I: O.K. What can you say about that, anything?
K: It's backwards! [He changed it to $4 + 3 = \square$] You can't go, 7 equals 3 plus 4."

(ibid, p. 84)

Kenneth's opportunity to "see" the relational logic of equality was thought to be missed by an apparently limited ability to see how the reversal of "$4 + 3 = \square$" is also true. Taken as a truth of mathematical equality, this expression and the questions asked about it highlight how Kenneth does not seem to have a natural intuition about the relationship of equality. Beyond that, the problem provides a way of organizing what was seen as Kenneth's individual understanding—a lens into how he sees, thinks about, and interprets the mathematics. But this is not simply "his," "natural," or "intuitive." It is a cultural distinction that identifies abilities that are given as Kenneth's own.

Part of a broader effort in math education research to identify individual abilities, skills, and challenges in learning math, this signification of individuality related math in schools to the logic of equality underlying "excellence in math for all." Understanding what were seen as the child's unique traits and differences was considered necessary to think about producing equal opportunities for everyone to learn and excel in math. Yet this individuality could not be understood without reference to a representation of all.

So, whereas the principles of identity and difference that expressed individuality for Johnny and Kenneth were not necessarily the same, the fundamental notions of interest, motivation, and intuition would establish equivalence between them. Similar to how the terms of equality could be reordered and seen as identical ($\square = 3 + 4$ to $4 + 3 = \square$), Kenneth and Johnny could be seen as distinct individuals—yet equal in their differences.

With individual motivations and intuitions as the reason of the back-to-basics reform, the cultural principles of identity and difference reorganized the curriculum and the child as a "naturally" but differently motivated individual. This ordering is visible at the intersection of research in the psychology of individual math learning and basic math skills that served to identify a child's natural motivations, abilities, skills, and interests as the basis of excellence in math—for all.

Motivation as the Equalizing Factor for Opportunity and Access

Excellence for all was coupled with a concern with how mathematics education was not meeting the basic requirement of providing all children with the skills to thrive as productive members of society. Further distinguishing notions of citizenship embedded in the previous new math reforms, questions about motivation to learn math continued to assemble with a concern for equality of opportunity. This opportunity was no longer granted by virtue of being seen as a citizen worthy of this right. In fact, because equal opportunities for all children to learn math could not technically be guaranteed, new notions of equality were thought to be measurable as the effect of an "equality of optimum motivation" (Nicholls, 1979). It seemed fairly straightforward that "we can say that someone is not developing his or her potential in, for example, mathematics, if that person is not optimally motivated to learn mathematics" (ibid, p. 1072). With equal access and opportunities now pinned to individual differences in motivation, it became common sense that a basic math curriculum was organized around what were taken as the motivations of each individual student.

Self-Discovery: Centering the Child

The idea that a meaningful and motivating mathematics started with each child became an organizing principle of the basic curriculum. By starting with the individual, rather than the mathematics, math pedagogy was to "encourage young children to respond to situations using their own ideas, language patterns; deemphasize the use of formal vocabulary" (Price Troutman, 1973, p. 427). This notion about the child as an individual, defined by his "own" ideas and language, embodied a way of seeing this individuality and making visible what its expression entailed.

This can be seen in a discussion of "Child-created Mathematics" (Cochran, Barson, & Davis, 1970). In solving the open equation $\square - 3 = \triangle$, a child named Leslie chose numbers to create equivalence between $\square - 3$ and $\triangle$ on either side of the equal sign. Mathematically, this depended upon substituting the $\square$ and $\triangle$ with chosen numbers to identify the unknowns with known quantities. The child was to make an equivalence as the comparison between the two unknown quantities. As part of this, the child produced a graph by translating the numbers and shapes to points on a grid. Through this "creation" (selecting numbers and plotting points), the shapes were given numerical identity—represented as a fixed quantity.

As a basic math skill, the open expression was to provide an opportunity for the child to create a mathematical understanding of equality as a relationship of like/unlike terms. Identifying the mathematical objects and producing the graph appeared to organize a way for Leslie to "create" an individual understanding. Apparently not satisfied with the way the graph of $\square - 3 = \triangle$ looked, Leslie expressed a desire for the graph to be more "straight up." Rather than correct this language use (by introducing the term "vertical"), the teacher was to provide Leslie with the opportunity to follow the intuition or hunch that the $\square$ and $\triangle$ were related and were not the same. Leslie appeared to make choices about the language and task direction while also discovering that changing the $\square$ would change the $\triangle$. This discovery was to inscribe and allow the child to internalize the feeling of being more satisfied with and motivated by the creation of mathematical understanding.

This "child-created" mathematics intersected with cultural norms of motivation expressed in a psychological language of a child's interests, abilities, and feelings of ownership. Placed at the center, the child was to be motivated to learn. Yet this required seeing each child as a distinct individual who was motivated in different ways.

As a way to define identities and distinctions among children, research emerged examining how individuals thought about and learned mathematics differently. One psychological discussion of math ability invited new ways to classify "the gifted or less able child" (Inskeep, 1970, p. 195). Within this range of ability, "mathematical precociousness" appeared as a psychological category to identify the child as mathematically talented beyond what was given as a normal measure.

The traits attributed to the mathematically precocious child were articulated in the language of interests, values, and personality. In a study of *Mathematical Talent: Discovery, Description, and Development* (Stanley, Keating, & Fox, 1974), the mathematically precocious child was characterized as a person who liked finding out things, discovering things, and learning things. This child also appeared to have a positive attitude toward and a general likeness of math (ibid). Together these traits explained success in school math through an individual interest in math and a discovery-based character developed through this interest. Within this identification of ability was a way for children to "see themselves" in their math classes. This seeing was more than a reflection of who the child "naturally" was. It was an invention of an individual—motivated by self-discovery, creation, and feelings of personal satisfaction with mathematics.

Creating one's "own" mathematics was thought of as an opportunity for a child to enact what were given as mathematical interests and intuitions. It was also to provide a way of to feel personally fulfilled, express an unbound sense of self as a motivated individual, and excel by discovering the basics of mathematics that were presumed to be there for discovery. Although not every child would achieve through self-discovery and creation. Given the assumption that "children's self-concept is usually bound up with their intellectual achievements," it became increasingly part of the common sense to identify individual strengths as well as weaknesses in learning math (Ginsburg, 1977, p. 130). As factors impacting achievement, an understanding of individual strengths and weaknesses were thought to provide all children with a better self-concept and an equal opportunity to be motivated to learn.

Informally interviewing children with "severe difficulties in the learning of arithmetic" became a way to "discover" both strengths and weaknesses as though they were already there. In solving $21 - 5 = 24$ a child named Bob appeared to use an incorrect strategy. The research emphasized Bob's self-realization that "Oh, I think I added" (ibid, p. 134). The idea that Bob came to his own awareness was to be understood as having an intuitive sense that he had "formally" done something wrong. Presumably, if Bob could discover for himself that he made a mistake in translating his "informal" mathematical logic, then he could eventually find a way to establish equality between $21 - 5$ and its equivalent term. Until then, being aware of himself would be a factor motivating future success. This awareness seemed to require the distinction between a formal, correct form of knowledge and a more natural, unofficial one. The difficulties in learning arithmetic, then, were tied to a child's presumably instinctual way of thinking about mathematical expressions.

Seen from the perspective of *The Psychology of Learning Mathematics* (Skemp, 1971), Bob's self-awareness was key to learning new mathematical concepts. Each person was thought to have "his own conceptual system," yet "the actual formation of a conceptual system is something which each individual has to do for himself" (ibid, pp. 28–29). Bob was to find his own way in mathematics through psychological notions of motivation. Ability mattered, but it was thought to be more important that each child put himself at the center of the mathematics, so even individual differences in ability, needs, and interests would meet the demand for all children to excel in mathematics.

As the equal sign connected to notions of self-discovery and self-creation, it embodied a cultural way of identifying the child as an individual motivated to learn mathematics. This relation of mathematical logic to cultural principles of identity also functioned to produce a notion of difference. Whereas certain kinds of kids were presumed capable of creating, discovering, and expressing their own basic and inherent mathematical ideas, others were not. This distinction was understood in terms of the psychological promise of "learning by discovery" (Gardner Thompson, 1973, p. 344). In solving the following series of problems:

$$1 + 1 =$$
$$1 + 1 + 1 =$$
$$1 + 1 + 1 + 1 =$$

"Caesar didn't seem to grasp that we were going in sequence and just adding one more each time" (ibid, p. 344). Not simply about one child, Caesar, this apparent inability to intuitively "grasp" the pattern expressed an apparent difference in ability. Given that "Caesar didn't seem to realize that problems like $8 + 1 = \square$ and $1 + 8 = \square$ were the same," it became possible to identify a presumed lack of self-awareness. This also rationalized a need for continued practice as the pedagogical approach to avoid the problem centered in a perceived lack of awareness and self-discovery.

Establishing equivalence between like problems provided a way to produce distinctions between the kind of child who appeared to take on the traits of self-discovering, -motivating, and -realizing and the child who could not. Here Caesar represented a child classified by difference in reference to the psychological norms of self-discovery, given visibility in the practices of learning mathematics. As more than mathematical rules, the principles of identity and difference embodied in the use of the equal sign also differentiated children based on who appeared to learn by discovery and could be motivated through a sense of self-realization.

Interests, Choice, and Self-Control: Games of Winners and Losers

As a way to provide equal opportunities for children to be motivated to learn math, certain practices of math pedagogy were to psychologically (re)order children as self-directed individuals. Beyond the notions of self-discovery already discussed, a motivated kind of individual was inscribed in the reorganization of the basic curriculum, expressed through the social and cognitive psychological principles of interests, choices, and self-control.

One commonsense way to motivate math learning was to appeal to children's interests. Through this lens, it seemed that "fostering enthusiasm through child-created games" motivated learning for children of all abilities (Price Troutman, 1973, p. 428). Following the assumption that "children are highly motivated to explain games clearly to others" (Golden, 1970, p. 114), math games were thought to provide children with the opportunity to express mathematical ideas and encourage interest in math.

Organized according to the logic that games provided a pleasurable medium through which to explore basic mathematical skills, they also embodied rules that ordered how the child was to play the game and enact that pleasure. That is, math curriculum games were not to be enjoyed at the expense of learning. The game's role in developing basic understanding was articulated in approaches to "[M]astering the basic math facts with dice" (Gosman, 1973). Here the dice and the mathematical expressions they represented appeared "loaded with activities to help children master basic number facts and enjoy the experience" (p. 330). In one turn of "[m]ultiplication-addition double trouble," children were to multiply the number on the dice by 5, following the formula $5 \times \#$ on the dice = total score.

On one level, solving equations like this ordered children's understanding of how to establish basic equivalencies between numbers given in multiple forms. On another, the equal sign carried rules for how to reason about and construct basic equivalences among children as a mode of comparison. To play the game the child was to create a relationship between a die, a numerical representation, a series of values added together, a total score, and winning or losing.

This mode of thinking was not just about mathematics—it was also about making choices within the cultural system of a "game" in which self-control and choice were to determine the outcome. Like mastering the math skills by following a formulaic method of multiplying by 5, enjoying the experience of the game was not to be left to chance. Students were to enjoy that the game was "completely under the control of the children. They must roll the dice, keep track of their scores, and decide to pass or play after each throw of the dice" (ibid, p. 331). This required children to make choices at every turn. How they made a choice was thought to impact not only the outcome of the game but also the extent to which the child enjoyed it as a mode of participating and being interested in math.

Whereas all kids could presumably learn math from and enjoy the game, only some of them would win. Nevertheless, math games were thought to "guarantee participants an experience that is like the experience they would have in the real world" (Cruickshank & Telfer, 1980, p. 77). Taking turns, calculating outcomes, keeping score, and deciding what to do next were to give all children the opportunities to "make decisions and live with the consequences" (ibid). In this way, solving $5 \times \#$ on the dice = total score was not only about following basic rules of mathematical logic. It was linked to a cultural expression of identity wherein children were to embody norms for personal decision-making and self-control. The calculation of the total score became a way to identify, compare, and differentiate children as winners or not.

Winning or losing was not necessarily a random or natural outcome. It was seen as a factor of one's calculated decision-making, interest, and effort. Thought of as the opportunity, "for participants to solve difficult problems themselves rather than to observe the way someone else solves them" (ibid, p. 76), the math game appeared to model what a child would encounter in the "real" world. Each child would need to decide to play the game and take advantage of each turn. Together "these exercises may give pupils a greater sense of control over their future" (ibid, p. 79). This sense of control was determined by how one played the game and what choices were made to impact the future outcome.

With the game determined by how children were identified by a numerical calculation, the total score became more than that. Multiplying 5 by the # on the dice became a measure of personal interest in the game, decision-making ability, and sense of control over one's life. In relating basic math facts to cultural principles for how children were to make their own choices and control their own futures, the equal sign used in the formula for playing embodied rules for living as a self-regulating and motivated individual. Organized by a particular way of reasoning about equality, the mathematical rules of the game assembled with cultural rules to produce equivalence as a mode of comparing things that otherwise seemed incomparable—children's self-control, decision making, and interest in playing by the rules.

Whereas losing the game could all be attributed to an unequal score, it could also be seen as child's lack of control over the game or inability to make decisions about basic math skills. If a child was not using them to play the game, make decisions, and win, then motivating this child to learn the basic skills was taken as a psychological concern about redirecting individual thoughts and actions.

The Psychology of Learning Mathematics: Managing Difference and Remaking the Unmotivated Child

In relation to the broader cultural logic of in/equality, unequal results and failure in math were taken to be the result of and therefore remedied through an understanding of individual differences. An emerging focus in the field, *The Psychology of Mathematical Abilities in Schoolchildren* (Kilpatrick & Wirszup, 1976) considered how "other conditions being equal, identical exercises and identical teaching methods yield essentially different results, these differences can be explained by differences in the pupils' abilities" (p. 5). Despite the hope that all children might be motivated to learn math, it appeared that "even with perfect teaching, individual differences in mathematical abilities will always occur—some will be more able, others less. Equality will never be achieved in this respect" (ibid, pp. 6–7). Individual differences in children's math learning abilities were deemed problematic and worthy of study as a way to make math excellent and equal for all children.

Addressing individual difference in math through the psychological lens of ability was part of making mathematical excellence for all. Yet this also entailed a "double gesture" (Popkewitz, 2008) as differences were produced as a pedagogical problem in the making of equality. This section will explore how emerging classifications of mathematical disabilities and apparent problems with motivation both expressed and produced notions of differences that were taken as the reason for why some students would not excel in math. It focuses on how the cultural principles of self-control, individual interests, and choice embedded in the promise of motivation served to identify abilities and produce differences as factors of individuality. In this, the section highlights how school mathematics was to be organized so that some children could make a new form of the self through math.

This focus on understanding individual differences during the back-to-basics movement coincided with new ways of determining when certain students were not granted equal opportunities. Distinctions in the opportunities in and access to mathematics (as well as other subject areas) were reported as inequalities. One such report, the Equality of Educational Opportunity Study (Coleman et al., 1966), became an important policy text. The study, also referred to as the Coleman Report, was requested by the Civil Rights Act of 1964 and aimed to document the availability of equal educational opportunities, specifically for minority groups.

Within a broader way of reasoning about inequality as the problem of difference, the report highlighted statistical categories that operated to reinscribe a common sense of failure among "Negros, Puerto Ricans, Mexican-Americans, Oriental-Americans, and American Indians." Given as distinctions from normalized classifications of equal opportunities, unequal educational opportunities were understood through these racial categories and marked as differences in "student characteristics, achievement, and motivation" (ibid). The incitement of this cultural truth of inequality was given intelligibility through a way of reasoning about difference as a problem of identifying the individual who was failing and prescribing ways to remake this child through schooling.

In the context of heightened interests in civil rights and making equal opportunities, mathematics education and research during back-to-basics reforms focused on how to understand differences in math abilities while trying to raise the standards of excellence for all students. The attention to individual differences was clear in an editorial note in the *Arithmetic Teacher* as an invitation to teachers and researchers: "Let us learn more about special learners and their problems" (Inskeep, 1970, p. 195). "Special learners" was inserted as a category to think about a seemingly vast range of human abilities—given qualities as either "the gifted or the less able child" (ibid).

Within this call, research was devoted to identifying what would be considered mathematical talent as well as a distinction of mathematical disability.

The fabrication of the child as either gifted or less able served as an explanatory factor of mathematical excellence. This distinction, like that of the "mathematically precocious child" discussed earlier, was given further nuance through measurement tools, like the California Psychological Inventory (Gough, 1969), to understand and define mathematical ability in psychological terms. The inventory appeared to give a "good indication of the general social functioning of the individual" (Weiss, Haier, & Keating, 1974, p. 128). Measuring factors such as sociability, self-acceptance, well-being, responsibility, and self-control, the inventory was used to characterize the traits of mathematically gifted boys. As a group, their achievement in math was attributed to internalized traits as flexible, independent, and self-actualizing people (ibid, p. 137). Boys who were deemed good at math were also given traits as the kind of individuals who could achieve to high levels as a result of their own self-guided efforts to be social, responsible, and in control of their learning and living.

At the same time that an apparent difference between mathematical ability and disability emerged, so too did a distinction between boys' and girls' achievement in mathematics. Gendered attributes, like those listed here, ascribed to notions of mathematical talent were circulated in studies that sought to understand the "unexpected and disconcerting" divisions in mathematical precocity drawn along gendered lines (Astin, 1974). Characterized as more sympathetic, tender, conscientious, and shy, girls were apparently not like boys, and their differences seemed to account for disparate levels of achievement.

Appearing to not exhibit as much self-acceptance, sociability, or self-control, girls were taken as less able to have the traits that were given equivalence with mathematical talent. Reporting how "the parents of girls have rather average or low expectations for their daughters, considering that these girls are exceptional with respect to mathematical aptitudes," expressed how it was the exception, rather than the rule, that a girl could excel at math (ibid, p. 84). The reportedly lowered expectations were not only their parents' but were a cultural expectation about how girls were to perform in math. Psychological traits ascribed as a girl's individuality were taken as the explanation as to why she would not necessarily be expected to achieve to the level of boys who were ascribed the traits of mathematical talent.

Whereas researchers in the field were trying to understand differences in mathematical abilities as the reason for unequal opportunities, "everywhere teachers are searching for ways to reexamine their understanding of mathematics teaching and learning as they work to resolve the problems of the inner-city child who is mathematically disabled" (Keiffer & Greenholz, 1970, p. 588). This distinction of disability, given as the characterization of a child's apparent lack of interest, initiative, and direction, was articulated as a geographical distinction in reference to the "inner city." But it was much more than that. Emerging to give expression to notions of differences as a way of reasoning about inequality, the inner city represented the child who was already outside of the boundaries of *excellence in math—for all.* This characterization was a finer distinction of the "disadvantaged child" that emerged in earlier new math reforms, drawn along geographic lines that encoded a racial distinction, made visible in terms of who appeared to be able, or not, to excel in math.

The child who was made up as irresponsible, disinterested, and unmotivated—as different from one that appeared to be motivated and mathematically able—needed a new sense of self, one that could be remade through mathematics. The common sense was that "what is taught is less important than how it is taught. It matters greatly that each student experience sufficient success to strengthen his confidence and pride in himself, to improve his self-esteem, and to encourage him to exert effort" (ibid, p. 590). In this case, it seemed that the traits of this individual got in the way of, rather than contributed to, math learning.

This kind of child, as mathematically disabled, was categorized through qualities distinguished from the normalized traits of motivation and interest in learning math. Characterized by a lack of effort, notions of disinterest were attributed to "problems of motivation, low reading and listening level" as well as a familiarity with "strife, discord, and rejection" that characterized him [sic] as an "expert at tuning out" (ibid, p. 588). So, in working to develop this child's self-image, the teacher could presume that "there is no profit for him in trying so no point in exerting effort" (ibid, p. 589). With the problem of difference identified in the child, the solution was given as teaching with "gentleness, respect, and understanding" so that "the fear of failure will be diminished, and the student will realize that his achievement is directly related to his own effort" (ibid, pp. 594–595). This had less to do with learning math and more to do with how to remake the child who did not fit the description of motivated, self-directed, and interested in mathematics.

Circulating within a notion meritocracy as the means to equal opportunities in school mathematics, this child was to work hard and identify with the goals of becoming more interested, self-respecting, and controlled. If the so-called inner city, mathematically disabled child put in effort, failure could presumably be avoided. Via the merits of motivation and "grit," the child could be remade into the kind of individual who could excel in math (Arithmetic Teacher, 1970, p. 4). As a term that has come into the present ways of examining student success in the United States, "grit" here was given as a psychological trait indicating one's ability to sustain interest, put in effort, remain focused, and regulate behavior. More than a psychological trait, it embodies cultural practices for organizing difference within the project of remaking a progressive, inclusive, and egalitarian society. "Grit," or lack thereof, rationalized how the problem of motivation placed in certain children required them to navigate inequalities and access opportunities by remaking who they were in comparison to who they should be.

During the back-to-basics reforms, math education research aimed at understanding individual differences had little to do with basic mathematical reasoning. The psychological principles of self-discovery, interest, control, and choice assembled with the math curriculum to organize who the individual child was and should be as a motivated kind of self. In this reorganization of the curriculum, motivation and its psychological characteristics served as the norm by which to identify equal opportunities for success in school math. These characteristics served as standards of equivalence by which to differentiate the individual who was to be seen as failing in relation to the group of *all* children.

The shift in the curriculum reforms, then, can be understood within the purview that mathematics education became a tool to make the child a self-sufficient and self-responsible individual, relying upon his or her own merits and motivations to access opportunity and success. At the same time, it also worked to identify the child who was deemed a failure and a threat to equal opportunity and the social cohesion this equality assumed.

A Nation at Risk: The Threat of Difference

The incitement of a cultural truth of in/equality in school math achievement during the 1970s brought with it new ways of organizing notions of failure and success as the differences among children. This chapter has explored this cultural logic of in/equality by relating the inscriptions of difference in the curriculum to the making and remaking of children as certain kinds of people. Arguing that the emphasis on the individual in the math curriculum was related to the problem of how to measure and understand the differences in children that constituted the commonsense notion of inequality, it has become evident how the rules for reorganizing math around notions of motivation to achieve "excellence in math—for all" also served to produce distinctions and reorganize the child who was not seen as that kind of individual.

The self-reliant and self-controlled child was made up as an individual who could be motivated toward personal success in mathematics. For this kind of student, mathematics was a tool by which to discover one self, pursue one's interests, and make personal choices. Organized according to the promise of motivation, the back-to-basics curriculum attributed these individual traits as the merits of success. Within this aim, the logic of focusing instruction on the individual rested on the premise that all children, although different in their interests, needs, and strengths, had an inherent ability for learning basic math skills.

This was about more than mathematics. The back-to-basics curriculum was reorganized to carry the promise of equalizing one's educational opportunity. Yet both the promise of equality and the problem of inequality seemed to be situated within the child. The child who was seen as failing, who did not have equal access to mathematics, was characterized as unmotivated, mathematically disabled, out of control, or disinterested or simply did not find him- or her- "self" in mathematics. For this child, mathematics was a tool to reshape the self through extra practice, effort, and grit, thereby solving what was given as the problem of inequality made visible through the inscription of difference. This problem of inequality in opportunity and its solution seemed to rest within the child as a natural instantiation of his "own" abilities, motivations, and interests.

The visibility given to a perceived lack of achievement and motivation attributed to certain children circulated and became a way of reasoning about how the nation was deemed "at risk" (National Commission on Excellence in Education [NCEE], 1983). In working toward a revision of mathematics education once again, the notion of risk was tied up with a continued hope for excellence. Math would continue to be considered one of the

> tools for developing their individual powers of mind and spirit to the utmost. This promise means that all children by virtue of their own efforts, competently guided, can hope to attain the mature and informed judgment needed to secure gainful employment, and to manage their own lives, thereby serving not only their own interests but also the progress of society itself.
>
> (ibid, p. 11)

The child was to become something else through mathematics education—a mature, secure, informed, managed, and self- and society-serving individual. This cultural statement about who the child is and should be as an individual through school math was distinct from the inscriptions of an independent, intelligent, and creative citizen embedded in the previous reforms. In the continuing focus to make *math for all* students, new ways of thinking about the relationship between children's math education and their identities as kinds of people would emerge and be given a more nuanced distinction. Some would continue to stand as the hope expressed in the previous quote, whereas others would be marked as the threat of difference and the reason for the nation's risk.

Note

1. The use of his and he throughout is to call attention to the gendered ways in which elementary math research was conducted and represented, by and large. With some exception, discussions of children's mathematics learning during this period were male oriented as indicated through the use of individual's names that appeared in the original text.

References

Arithmetic Teacher (1969).16(6)
Astin. (1974). Sex differences in mathematical and scientific precocity. In Stanley, J.C., Keating, D.P., & Fox, L.H. (Eds.), *Mathematical talent: Discovery,*

description, and development. Baltimore, MD: The Johns Hopkins University Press. pp. 70–86.

Bellah, R.N., Madsen, R., Sullivan, W.M., Swidler, A., & Tipton, S.M. (1985). *Habits of the heart: Individualism and commitment in American life.* Los Angeles, CA: University of California Press.

Borstelmann, T. (2011). *The 1970s: A new global history from civil rights to economic inequality.* Princeton, NJ: Princeton University Press.

Brodinsky, B. (1977). Back to the basics: The movement and its meaning. *The Phi Delta Kappan, 58*(7), 522–527.

Cochran, B.S., Barson, A., & Davis, R.B. (1970). Child-created mathematics. *Arithmetic Teacher,* March, pp. 211–215.

Coleman, J.S., Campbell, E.Q., Hobson, C.J., McPartland, F., Mood, A.M., Weinfeld, F.D., & York, R.L. (1966). *Equality of educational opportunity.* Washington, DC: National Center for Educational Statistics—U.S. Department of Health, Education, and Welfare.

Cruickshank, D.R. & Telfer, R. (1980). Classroom games and simulations. *Theory into Practice, 19*(1), 75–80.

Gardner Thompson, M. (1973). Hidden implications for change. *Arithmetic Teacher,* May, pp. 343–349.

Ginsburg, H.P. (1977). *Children's arithmetic: The learning process.* New York, NY: Van Nostrand Reinhold, Inc.

Golden, S.R. (1970). Fostering enthusiasm through child-created games. *Arithmetic Teacher,* February, pp. 111–115.

Gosman, H.Y. (1973). Mastering the basic facts with dice. *Arithmetic Teacher,* May, pp. 330–331.

Gough, H.G. (1969). *Manual for the California psychological inventory.* Palo Alto, CA: Consulting Psychologists Press.

Inskeep, J.E. (1970). As we read: Editorial comment. *Arithmetic Teacher,* March, pp. 193–195.

Keiffer, M. & Greenholz, S. (1970). Don't underestimate the inner city child. *Arithmetic Teacher,* November, pp. 587–595.

Kline, M. (1973). *Why Johnny can't add: The failure of new math.* New York, NY: St. Martin's Press.

Krutetskii, V.A. (1976). *The psychology of mathematical abilities in schoolchildren.* (J. Kilpatrick & I. Wirszup [Eds.], J. Teller, Trans.). Chicago, IL: University of Chicago Press.

National Commission on Excellence in Education (NCEE). (1983). *A nation at risk: The imperative for educational reform: A report to the nation and the secretary of education, United States department of education.* Washington, DC: The Commission: [Supt. of Docs., U.S. G.P.O distributor].

National Council of Teachers of Mathematics (NCTM). (2000). *Principles and standards for school mathematics.* Reston, VA: NCTM.

Nicholls, J.G. (1979). Quality and equality in intellectual development: The role of motivation in education. *American Psychologist, 34*(11), 1071–1084.

Popkewitz, T.S. (2008). *Cosmopolitanism and the age of school reform: Science, education and making society by making the child.* New York, NY: Routledge.

Price Troutman, A. (1973). Strategies for teaching elementary school mathematics. *Arithmetic Teacher,* October, pp. 425–436.

Skemp, R. (1971). *The psychology of learning math.* New York, NY: Penguin Books.

Stanley, J.C., Keating, D.P., & Fox, L.H. (Eds.) (1974), *Mathematical talent: Discovery, description, and development.* Baltimore, MD: The Johns Hopkins University Press.

Turner, R. (1976). The real self: From institution to impulse. *American Journal of Sociology, 81,* 989–1016.

Weiss, D., Haier, R.J., & Keating, D.P. (1974). Personality characteristics of mathematically precocious boys. In Stanley, J.C., Keating, D.P., & Fox, L.H. (Eds.), *Mathematical talent: Discovery, description, and development*. Baltimore, MD: The Johns Hopkins University Press. pp. 126–139.

Zurcher, L.A. (1977). *The mutable self: A self-concept for social change*. Beverly Hills, CA: Sage Publications.

6 Planning the 21st-Century Future

Standardizing Mathematical Kinds of People to Manage Risk

> It is students' acts of construction and invention that build their mathematical power and enable them to solve problems they have never seen before.
>
> (National Research Council [NRC], 1989, p. 59)

This statement was released to the United States public in a document titled *Everybody Counts: Report to the Nation on the Future of Mathematics Education* (NRC, 1989). The National Academy of Sciences, the National Academy of Engineering, and the Institute of Medicine, acting through the National Research Council, put out this urgent call to "revitalize" and "rebuild" mathematics education. Looking ahead to the year 2000, it seemed evident that "to participate fully in the world of the future, America must tap the power of mathematics" (ibid, p. 1). As a common bond that all people seemed to share, the notion of "mathematical power" intersected with cultural notions of identity and equality in ways that ordered who was/was not part of *everybody*. This chapter focuses on how, during the standards-based reforms of the 1980s in the United States, the inscriptions of identity and difference were re-signified and set the standard for new ways of reasoning about who children are and should be through elementary school mathematics.

The norms of identity that were to include all children into *math for all* shifted from the ideals of citizenship in the new math reforms, to distinctions of individuality during a back-to-basics movement, to standards of personhood during the standards-based reforms of the 1980s.

This chapter examines this shift by focusing on how the principles and practices of equality embedded in the *Curriculum and Evaluation Standards for Mathematics* (National Council of Teachers of Mathematics [NCTM], 1989) organized the standards reform in the United States. Arguing that the math curriculum was about more than setting standard expectations for math teaching and learning, it explores how elementary math education intersected with psychological principles of identity and difference to organize children as *problem-solving* kinds of people with *mathematical powers* who could solve the problems of everyday life.

Articulated through the rules of cognitive and social psychologies that fabricated all people as social beings able to think about and express their thinking, this ordering of identity, equivalence, and difference made it reasonable to think that standards would "ensure that *all students benefit equally* from the opportunities provided by mathematics" (NRC, p. 89, my italics). *All children* were thought capable of learning mathematics because math was about problem-solving—something all people could presumably do. This way of thinking about equality and the equivalence given to children are explored in and through the use of the equal sign in standard math problems. More than about mathematics, it argues how the use of the equal sign organized notions of math power, problem-solving, and mathematical literacy as the social and psychological standards of personhood.

The common sense that *everybody counts* implied a cultural way of reasoning about an *equal benefit* inscribed in *math for all*. Marking a shift from a focus on individual differences in math learning that characterized a back-to-basics movement, the standards period of reform was to be organized according to a generalized notion of everybody. In the previous reforms, the individualization expressed a cultural thesis about a motivated, self-controlled, and self-directed child.

The apparent need to "tap the power of mathematics" (NRC, 1989, p. 1) during the standards-based reformed embodied a new way of thinking about who children are and should be through mathematics that gave further distinction to who was, and who was not, part of *math for all*. This chapter highlights research in mathematics education and changing psychologies that emphasized the awareness and communication of one's thinking assembled to give characteristics to the child's development of a mathematical form of power while also producing mathematical illiteracy as a problem of difference and inequality.

(Re)Setting the Standards: The Mathematical Power of Problem-Solving

During the movement into standards-based reforms, mathematical thinking was presumed as something that everybody did. This can be seen in emerging discussions of "logical-mathematical" kinds of intelligence that were thought to frame one of many, yet universal, ways of thinking about and seeing the world (Gardner, 1983). With a natural capability to think logically and mathematically, all people could be seen as having a particular power to see and reason in mathematical ways. Even more, it was thought that everybody was a mathematician and does mathematics consciously (NCTM, 1989; NRC, 1989). School mathematics, then, was to "endow all students with a realization that doing mathematics is a common human activity" (NCTM, 1989, p. 6). The naturalness and rationality given to mathematical thinking, as a capacity of mind common to everybody, established equivalence among *all* people.

Mathematical modes of thought were presumed to be cultural practices that bound people together across both time and space. This belief was expressed in how the ancient Greeks appeared to use math and its powers as a logical way to see, think about, and order a complex and chaotic world (Johnson, 1983, p. 19). Articulated in this example of the logic of equality, the abstraction of mathematical power and its practicality in problem-solving was represented as a universal and symbolic way of ordering the world according to predictable patterns.

Consider the use of the equal sign in "a pattern such as the following":

$$2 + 3 = 3 + 2$$
$$9 + 5 = 5 + 9$$
$$27 + 58 = 58 + 27$$
$$132 + 6 = 6 + 132$$

(ibid, p. 20)

This pattern, represented as a series of whole number sums, seemed to be generalizable as a logical rule. "For counting numbers, called n and m here, it is always true that $n + m = m + n$" (ibid, p. 21). No matter the order, two discrete things can be added together to establish an equal identity. This symbolic expression of equivalence assumed an "If . . . then" rule: If $2 + 3 = 5$ and $5 = 2 + 3$, then it follows that $2 + 3 = 3 + 2$. Accepted as a truth of mathematical logic, this expression of equivalence and identity inscribed a law of equality embodying a notion of the logical power of mathematics. This presumably universal mathematical logic intersected with a cultural standard by which to identify the power to think mathematically. This power relied upon what was considered logical thought, ordered by the ability to see patterns and generalize relationships among objects in the world. But it was about more than that.

Mathematical power inscribed a mode of living in the world. Articulated in *The World Book of Math Power* (Johnson, 1983), mathematics was thought of as a "how-to-science, a friend and ally, a way of solving problems of daily life" (ibid, p. 492). Given new expression as math power, mathematical modes of thinking were to be used as a way to solve everyday problems in ways that reframed the expectations for school mathematics. Math's focus was no longer expressed as abstract reasoning or mastering basic skills. Problem-solving became the focus of school mathematics and reestablished the expectation for the kind of people children were to become.

To use math as a way to solve everyday problems, it was reasoned that "students need to view themselves as capable of growing mathematical power to make sense of new problem situations in the world around them" (NCTM, 1989, p. 6). The vision of students' capabilities to grow and use mathematical powers was expressed through the image of the child as a *problem solver*. The fabrication of this kind of person will be examined in the next three sections through a focus on how norms of communication, metacognition, and confidence were inscribed in the curriculum as standards for making *everybody count*.

Math as a Language and Tool for Becoming a Mathematical Problem Solver

As the standards were in development, *Mathematical Problem Solving* (Schoenfeld, 1985) was a topic of research emerging to examine the "spectrum of behaviors that comprise people's mathematical thinking" and "to explain what goes on in their heads as they engage in mathematical tasks" (p. 5). This psychological explanation, as a mode of establishing equivalence among people as "good problems solvers," relied upon a person's ability to communicate, describe, and justify their thought processes. The association between language and the representation of mathematical thinking framed the expectations by which children were to express themselves as a new kind of person, mathematical problem solvers.

The emphasis on language use and communication was not necessarily new to the standards period of reform in the 1980s. Vocabulary and its precise use were granted as integral to understanding foundational concepts of equivalence and equality during the new math curriculum reforms examined in Chapter 4. During the previous back-to-basics reforms, it was believed that "speech (which has to be learnt) is essential for the formation and use of the higher order concepts which, collectively, form our scientific and cultural heritage" (Skemp, 1971, p. 28). In Chapter 5, this learning was visible in the curriculum as a process in which children's "own" language was appropriate for creative self-expression in mathematics.

Verbal communication and language use were given new emphasis during the reorganization of math education in the 1980s. Mathematical problem-solving and power was defined through the use of mathematics as a language not only to *represent* a mode of thinking that was presumed to be already there but to *develop* it. In *Speaking Mathematically: Communication in Mathematics Classrooms* (Pimm, 1987), it was assumed that a large "part of learning mathematics is learning to speak like a mathematician" (p. 76). As a social behavior, speaking mathematically was given equivalence with thinking mathematically. In this equation, both mathematics and language learning were described as natural abilities, such that the "intellectual powers of children evidenced in the mastery of their first language" served as the standard by which to deem all children capable of expressing a presumably inherent mathematical power (ibid, pp. 197–198).

If all children had intuitive abilities to learn language *and* inherent mathematical-logical frames of thought, then it would follow that they could all be taught to "grow in their ability to communicate mathematically and use higher-level thinking processes" (NCTM, 1989, p. 23). Communication in general, and the use of a particularly mathematical language, appeared to nurture this growth.

This way of thinking about how children use mathematics as a language and mode of reasoning to communicate about and solve what were given as everyday problems gave meaning to the Standard #2 of the K–4 curriculum, *Mathematics as Communication*. This standard included an activity wherein children were to use language and counters to represent the problem $14 - 5 = \square$. Children were expected to work together in groups to model and discuss three problems. The emphasis was not necessarily on finding the correct answer. The modeling and discussion of the problems was so that they could "see how problems that appear to be different in fact share the same underlying structure" (ibid, p. 27). Through this problem-solving— evidenced by a representation of thinking through discussion—they were supposed to conclude that $14 - 5 = \square$ is a common way of representing a "problem" about Maria and some pencils, Eddie and several balloons, and Nina and Pedro's seashells.

The shared structure to the problems was apparently visible through a mathematical discussion of how they were different—to see how they were alike. This discussion was to lead to a consensus about what the problem was and how to represent it. Modeling with both the counters and the mathematical expression was related to an importance in the curriculum place on "translating a problem or idea in a new form" (ibid, p. 26). This translation involved: decoding the text, communicating to establish a consensus about what the problem is, modeling the problem with counters, and reorganizing the model into a symbolic representation to talk about its meaning.

Using mathematics as a mode of communication seemed to establish a common way for children to see and talk about problems in their world—given the apparent differences in their abilities. So, in spite of the various contexts in which the problems were given, children were to discuss the distinctions, reduce them, and agree to represent their equivalence as a general problem of difference. The relationship between 14, 5, and $\square$ in each situation was to be organized by the rules of subtraction and given as $14 - 5 = \square$. The problems were all alike in that they could be seen as subtraction, or a problem of difference. Embedded within the rules for establishing equality between things that did not seem to be alike but shared a similar organization was a particular way of ordering equivalence and a way to regulate difference.

This was not just in the mathematics but transmogrified into cultural modes of reasoning about how similarities could be organized as standards for children and their behaviors. As the practices of problem-solving intersected with the rules of cognitive and social psychologies, a new way of seeing children emerged. Identifying children as distinct in their abilities, yet presumably always learning through communication with others and the world, the rules of psychology assembled with the logic of mathematics to produce an image about who the mathematical problem solver was and should be. Rather than focusing on individuality and differences in learning, the child as a mathematical problem solver was articulated through principles of psychology that fabricated all people as social beings able to communicate mathematical thinking as a process.

On one hand, the value of math was justified in how it appeared as a universal communicative tool, useful for expressing, interpreting, and translating the world. On the other, the representation of math as a language gave equivalence to all children and their apparently natural capacity to learn and talk about mathematics. This way of thinking about the standard expectations for children invoked a commonsense way of thinking about who *everybody* is and how they come to count as part of the *all*.

Modeling with Math: Representing Problems and Metacognitive Awareness

Returning to the presumed "spectrum of behaviors that comprise people's mathematical thinking," the person who exhibited mathematical power was expected to self-monitor as a way to assess and control problem-solving as a process (Schoenfeld, 1985, p. 5). Within the concerns about standardizing problem-solving methods in mathematics, the awareness of one's thoughts established a cultural norm of the kinds of people who make good problem solvers—as another demonstration of mathematical power. "The issue is not what the students know, it is how they use that knowledge" that made them successful in solving problems in mathematics (ibid, p. 32). The "correct" use of mathematical knowledge became a particular trait of the mathematically powerful problem solver.

Articulated in a standard "process model" of problem-solving, *metacognitive awareness* emerged as a psychological characteristic to identify the right kind of problem-solving practices and person. Operating at a "meta level" were processes of planning, monitoring, and evaluating that people were supposed to be aware of to change intellectual behaviors, engage in problem-solving, and learn. Organized by rules of cognitive psychology, this metacognition in mathematical problem-solving was defined according to practices of self-monitoring, self-regulating, and evaluating cognitive activity (Silver, in Schoenfeld, 1987, p. 49). More than components of mathematical thinking, these were psychological traits that ascribed identities of kinds of people who were aware of and able to think about their thinking.

Given as traits of the good problem solver, meta-awareness and self-regulation ordered the thoughts and actions of children in their apparent ability to communicate mathematically. Related to the psychological norm of metacognition—as the guide to changing and transforming intellectual behaviors—even when kids seem to be wrong, what was "of primary importance is the value that children derive from reflecting on their responses" (NCTM, 1989, p. 27). A cultural norm of equivalence by which to determine difference is embedded in the statement about "the value that children derive." Within this evaluation, expressed as the ability to see, assess, communicate, and plan one's mathematical thinking was a distinction of the kinds of people who did not yet meet the standard.

In this image of the child as the child as a mathematical kind of person, the norms of problem-solving overlapped with cultural rules of how to identify children as able to find "value" in reflecting on their responses. When looking at the conversation that follows between a generalized teacher (T) and pupil (P) who could represent *everybody*, mathematical rules are instantiated as cultural expectations for how the child should model, communicate, and think about the problem 26 − 17 = □.

T: . . . [Y]ou start it off and tell me what you are doing.
P: *Put one there.*
T: No. Let's start from the very beginning. Six take away seven. Can you do it?
P: *No.*
T: No. Why can't you do it?
P: *Cos it's . . . bigger number on the bottom.*
T: All right. Because six is smaller than seven. All right? So what do we do then? We go to the . . .
P: *The . . . the units.*
T: No. What column's that? The tens column. Right. And what do we do there?
P: *We cross that one out . . . and then we put one there.*
T: We take a . . .
P: *Er . . . er . . . er . . .*
T: . . . What do we take from the tens column? We take a ten don't we? One ten.
P: *Put one there.*
T: Yes you've got one left there. And where do you put the ten you've taken?
P: *There.*
T: You put them in the units column, right. How many units have you got?
P: *Twenty-six.*
T: No. Put your ten in the units column. No. No. Come on, you go to the units column and you take a ten. Where do you put the ten? We put it in the . . . units column, don't we. Like we did there, and there, and there, and there.

(Pimm, 1987, pp. 65–66)

A cultural standard of how to become aware of, think about, and communicate one's thought process is embedded in this problem-solving and its discussion. The repetitive instantiation of "we" in statements such as "We take a . . ." and "What do we do now?" embodies normalized rules about how this problem should be thought about, communicated, modeled, and solved. The rules also carry an expression of what isn't being done. In ways that seem almost natural, the psychological theories about a mathematically powerful problem solver organize what children should know and be able to do in ways that distinguish who and who is not this child. Within this expression of mathematical power, the child here could be considered unsure and unable to represent—either in mathematical language or models—what was given as a mathematical problem-solving process.

Made into pedagogical problem, the interaction assumes the use of mathematics as a cultural language that is not merely that of mathematics. The student's general statements of "there" are rephrased into a "mathematical" language indicating "You put them in the ones column, right." It also carries the expectation that the student would have a clear plan for how to strategically solve the problem and where to start—"at the beginning." The repetition of "No" and questions about what to do next in response to the student's utterances of "*Er . . . er . . . er . . .*" express a sense of certainty in terms of how this problem should be done. At the end of the discussion, the assumption that the student would see a pattern and model it is articulated in the reminder "Like we did there, and there, and there, and there."

The logic of equivalence and identity in the problem $26 - 17 = \square$ embodied the cultural standard of people living as mathematical problem solvers through psychological principles of metacognition and communication. This mode of characterizing a cultural standard of problem-solving gave equivalence to all children—as having equal opportunities to think and speak mathematically to benefit from learning math. Children's participation as part of *everybody* was to be organized around the assumption that they could plan, assess, rethink, and communicate what were seen as their problem-solving powers. This power was given further distinction in how children carried out the problem-solving and how they felt about it.

Mathematics as Reasoning: Strategies for Making Sense and Building Confidence

In the discussion of Standard 3, Mathematics as Reasoning, children's math power was to emerge not only through learning math as a language and an awareness of their thought processes but also through an internalization of the beliefs that mathematics makes sense and that they can make sense of it. In fact, "a major goal of mathematics instruction is to help children develop the belief that they have the power to do mathematics and that they have control over their own success or failure" (NCTM, 1989, p. 29). The power for all to benefit equally from mathematics seemed to rely upon believing that math could be used as a form of reasoning to reliably solve any problem.

This idea that math makes sense carried a notion of how that sense was produced. Particular problems were to thought to "[help] children see that mathematics makes sense" (ibid). Solving $35 - 19 = \square$ was not simply about children computing the answer. This problem was posed in relation to the idea that "children need to know that being able to explain and justify their thinking is important and that how a problem is solved is just as important as its answer" (ibid). Although children might be able to identify the correct answer, the goal was to have a strategic and analytical way to think about the problem that was more than the answer. The justification emphasized here was about knowing how and why the answer was sensible.

A series of questions posed by the teacher were thought to illuminate an apparently inherent logic organizing how children should see and think about mathematical equivalence as reasonable but not necessarily self-evident:

"Do you think it would help to know that $35 - 20 = 15$?"
"How would it help to think of 19 as $15 + 4$?"
"Would it help to count on from 19 to 35?"

(ibid, p. 30)

These questions implied strategies for seeing the sense built into the related expressions of equality. To the first question, children were to respond, "Yes." Knowing that $35 - 20 = 15$ is a true and valid expression was supposed to help them think strategically about the problem $35 - 19 = \square$. By comparing the 19 to 20 as 1 less, children should reason that the 15 compared to the $\square$ would be 1 more. Reframed as a different, yet related problem, $35 - 20 = 15$ was thought to organize a reasonable way to see a mathematical relationship of equivalence between 35, 19, and $\square$.

This way of making sense of the expression appealed to an authority of the identity given to whole numbers and the relationships among them. The meaning ascribed to numbers meant that they could be seen as identical (such as 35 and 35) and different (like 20 and 19). As a series of ordered and predictable patterns, the identity given to numbers formed the relationship by which to determine such equivalence and difference. In the standards, the ability to distinguish among numbers, understand how they were related, and predict their patterns was referred to as "number sense" (ibid, p. 38). This sense of number entangled with the equal sign (=) was the basis on which children should make "valid arguments" about mathematical equivalence and nonequivalence.

Assigning identity to objects is not merely about mathematical thinking. The logic about establishing equivalence embodied in the equal sign resembled cultural principles of identity that ordered how children were to make valid arguments as a way to build confidence and a belief in their abilities to learn mathematics. "Number sense" was not only a way of thinking about numbers but was related to psychological notions of how that sense was built.

The discussion about $35 - 19 = \square$ was to help children see that "the ability to reason is a process that grows out of many experiences that convince children that mathematics makes sense" (NCTM, 1989, p. 31). This convincing was about showing children that math made sense so that they would believe in, be aware of, and appreciate it. The problem-solving was to convince children that math just made sense. Presumably, if children believed in and saw the power of mathematics in many situations, they could become more sure and certain that they had the power in them to know, do, and show it.

The inscription of mathematical power was to be assessed by "examin[ing] students' disposition toward mathematics, in particular their confidence in doing mathematics and the extent to which they value mathematics" (NCTM, 1989, p. 205). Taken as a psychological trait of people who engaged in what appeared as effective problem-solving, confidence was associated with mathematical power and became possible to evaluate. But this was not just an assessment. This notion of confidence, embodied in the problems children were to solve and the rules for solving, framed a standard way of reasoning about how all students should learn math because it was thought that this confidence could be built from the validity granted to math itself.

Determining whether or not a student was confident and valued math was a way to psychologically (re)order the feelings, perceptions, and beliefs that children should have as mathematically powerful kinds of people.

The mathematical notions of identity and equivalence embodied in expression like $35 - 19 = \Box$ interacted with cultural modes of representing the child as capable of using math as a language, representation of thought, and mode of reasoning to communicate beliefs in and control over the logic of mathematics. Organized by psychological principles identifying effective problem-solving strategies, children were seen as people who could make sense of, plan, rethink, and confidently communicate a well-organized, logical, and justifiable mathematical thought process. The shaping and reshaping of mathematical thought was tied to a reorganization of the child's value system as well as his or her value as a person.

The aim of the standards curriculum was to provide experiences for *all students* to value math, be confident in one's abilities, become a mathematical problem solver, and communicate and reason mathematically (NCTM, 1989, pp. 5–6). Elaborated in the previous three sections through an analysis of the standard use of equal sign, these aims were expressed in how children were to see, think, speak, represent, and communicate mathematically—as the expression of mathematical power. Given that "math power is a term important for everyone" (Johnson, 1983, p. 8), it framed the common sense about how *all* children should know and do mathematics to live in the world as a person with a mathematical form of power. These goals produced an image of the kind of person who appreciated math, respected its certainty, understood and spoke its language, and saw through its perspective to solve the problems of everyday life.

Mathematical Literacy as Power in the "Real World"

Marking a shift in emphasis from all children as creative citizens in the new math reforms to self-motivated individuals in the back-to-basics movement, the standards-based reforms organized and maintained that all children could learn math as expressive, self-aware, and confident problem solvers given the presumed mathematical powers inherent in all people. Whereas independence, creativity, motivation, and self-control still appeared as important psychological factors in learning math, they became reassembled as traits of a problem-solving kind of person who appreciated and utilized a power in mathematics.

Through the notion of mathematical power, it became reasonable to think that *everybody* could be a confident, expressive, self-aware *problem solver.* The standards, then, in working to build mathematical power in the child as a problem solver would "ensure that all students benefit equally from the opportunities provided by mathematics" (NCTM, 1989, p. 89). Once again, the goals of the reforms associated ways of thinking about learning mathematical equality with efforts to produce social equality. Importantly, the equal opportunities provided by mathematics were framed by "the intent of these goals is that students will become mathematically literate" (ibid, p. 6). As an effect of mathematical power, mathematical literacy became necessary to being the kind of person who could benefit from math in schools and from whom society could benefit.

The literacy and power embedded in mathematics articulated a desirable kind of person. The production of a "mathematically literate worker" and "mathematical literacy" assumed that "mathematical literacy is essential as a foundation for democracy in a technological age" (NRC, 1989, p. 8). This link between democracy, technology, and elementary math education was a historical one that emerged during the middle of the 20th century. Here this relationship shifted and reimagined the child as a *mathematically literate kind of person* who could act in the world in ways that seemed to draw upon a mode of thinking called "numeracy." The sense of mathematical literacy and numeracy as a way of living is evident in statements about how "with numeracy comes increased confidence for individuals to gain control over their lives and their jobs. Numeracy provides the ability to plan, to challenge, and to predict" from a series of choices and alternatives (Steen, 1990).

Emerging in a web of technological, democratic, and economic considerations that generated cultural principles about a kind of person, the mathematically literate were presumed capable of sustaining their own livelihoods while contributing to the "world of work in the twenty-first century" (NRC, 1989, p. 11). This world was taken to be "less manual but more mental; less mechanical but more electronic; less routine but more verbal; and less static but more varied" (ibid). Certain kinds of people were thought to be best "prepared to cope confidently with quantitative, scientific, and technological issues" both now and in the future (ibid, p. 12). Gaining a sense of control and confidence seemed to provide individuals with the tools of thought and action with which to plan the future by making confident decisions in the present. Improving mathematical literacy, or numeracy, inscribed a mathematical model of the world that was thought to make a better kind of person—and, in turn, produce a better kind of worker.

In the present and beyond the context of the United States, this common sense is reiterated in statements by the OECD (2009) about how mathematical literacy is an individual's capacity to identify and understand the role that mathematics plays in the world, to make well-founded judgments, and to engage mathematics in ways that meet the needs of that individual's current and future life as a constructive, concerned and reflective citizen.

(p. 14)

The mathematically literate appeared as the kind of people who are confident, in control, and could make reasonable choices now and for the future. As an expression of mathematical power, mathematical literacy was seen as an integral part of using math as a lens for seeing, thinking, acting, and communicating in and about the world.

Whereas the standards aimed to emphasize "connections between mathematics and the real world and [encourage] children to recognize and use a variety of situations and problem structures" (NCTM, 1989, p. 12), this instantiation of the "real world" carried an assumption of what was considered real in the world as well as what constituted a problem in it.

Mathematical Illiteracy as the Problem of Difference

During the standards-based reforms in the 1980s, mathematical literacy emerged as a common way of reasoning about the democratic identity of the nation, its citizens, its workforce, and its schoolchildren. In the equivalence given to mathematical literacy and power as the articulation of a new kind of person, a notion of difference appeared as the distinction from that standard. "Math power can mean the difference between confidence and insecurity; vitality and boredom; goals and regrets" (Johnson, 1983, p. 8).

The expression of this difference became visible as distinctions of insecurity, boredom, and regret in discussions of the problem of "innumeracy." Taken as "an inability to cope with common quantitative tasks," innumeracy and its consequences were apparently "magnified by the very insecurity that it creates" (Steen, 1990, p. 216). "Innumeracy" was associated with psychological notions of apprehension, fear, and uncertainty (ibid). It also assembled with concerns over a culturally and historically specific form of "math anxiety" that was defined as expressions of helplessness, guilt, unwillingness, uneasiness, apprehension, and boredom in mathematics (Chavez & Widmer, 1982; Cockcroft, 1982; Stodolsky, 1985).

As a problem in the pedagogy of mathematics, the differences that were produced to identify innumerate kinds of people as insecure, disengaged, ignorant, and unreflective were conceived within and emerged alongside the psychological traits that produced a mathematically powerful and literate kind of person.

Fabricated as a kind of person, the mathematically illiterate was described as having "a lack of numerical perspective, an exaggerated appreciation for meaningless coincidence, a credulous acceptance of pseudosciences, and an inability to recognize social trade-offs" (ibid, pp. 5–6). Apparently, this person could not appreciate the sense of certainty a mathematical perspective would lend to one's approach to living and problem-solving in the social world. To use another language, they were seemingly an insecure and passive group who left life to chance, could not assess benefit and risk, and were otherwise unplanned, unpredictable, and unreliable. Not merely about math, the distinction appeared in relation to the makeup of a kind of person who was desired—confident, meta-cognitively aware, and expressive of logical thoughts.

These differences were generated in relation to the cultural distinctions that seemed to threaten a democratic sense of equality and opportunity. That is, "apart from economics, the social and political consequences of mathematical illiteracy provide alarming signals for the survival of democracy in America" (NRC, 1989, p. 13). In *Innumeracy: Mathematical Illiteracy and Its Consequences* (Paulos, 1988), the social and political consequences of the "mathematically illiterate" were seen as the embodiment of various inadequacies, incompetence, and an overall lack of ability to function as a so-called productive member of society.

Seen as less productive and without the power to act in intentional, rational, and organized ways, this classification of difference was a shift from previous reforms. Given a new name, the markers of difference that defined the problems of inequality were no longer articulated as "disadvantaged," "disabled," or "inner-city" kinds of people as in the previous reforms and examined in the earlier chapters. Yet the shift to a "mathematically illiterate" kind of person carried the same comparative way of reasoning about difference and the problems of inequality it presumes.

During the standards-based reforms, mathematical illiteracy became visible in the relation to the problem that "far too many students, disproportionately minority, leave school without having acquired the mathematical power necessary for productive lives" (NRC, 1983, p. 73). This productive life was to be lived by the kind of person with an "acquired mathematical power"—a life that minority students were apparently not living.

Distinguished from the mathematically literate and powerful, and seen as unpredictable, dependent, unreliable, insecure, helpless, and unproductive—the mathematically illiterate divide was drawn along economic and racial lines. In its public expression, mathematical illiteracy was assigned to Hispanic and Black students. It was believed that "we are at risk of becoming a divided nation in which knowledge of mathematics supports a productive, technologically powerful elite while a dependent, semiliterate majority, disproportionately Hispanic and Black, find economic and political power beyond reach" (NRC, 1989, p. 14). As the representation of a "a nation divided both economically and racially by knowledge of mathematics" (ibid, p. 13), the mathematically illiterate was the cultural expression of difference and named poor, Black, and Hispanic students as the problem in the aim of ensuring all students benefit equally and live a productive life through mathematics. Although it would seem that everybody counts, some would count as risks to the nation's political, technological, economic, and cultural growth.

Mathematical power, as a way of representing the world and the mathematically il/literate person's place in it, gave meaning to what the productive kind of life looked like and who could live it. The productive life, lived as a mathematically powerful person, was not equally available to everybody and impacted "which doors are open and which are closed as students leave school and enter the world of work" (ibid, pp. 74–75). Embedded in the equivalence given to children as equally capable of possessing and demonstrating mathematical power was a comparative frame with which to see children who apparently could not draw upon a mathematical power to benefit from all that mathematics was to provide.

The inscription of math power as a mode of establishing equivalence among *all* students also produced a notion of difference as the reason for inequality. This distinction gave meaning to the common sense that "without common standards, different communities will move in different directions, inevitably widening the gap in mathematical power available to children raised and educated in different circumstances" (NCTM, 1989, p. 89). The "gap in mathematical power"—as an expression of difference—was given as the reason for the standards. The standards and their common expectations appeared to solve for the problem of difference given as the representation of certain communities, directions, and circumstances that seem to fall outside of the standard.

But mathematical power is not natural or something inherently part of children living a different kind of life. One's capacity to yield mathematical power and live a productive life through it was culturally produced in ways that made it some "thing" to be acquired through schooling—available for some and not for others. A gap in mathematical power, as something some had and others did not, was taken as the expression of inequality. Within this logic, the gap was economically and racially coded and stood to represent the kind of life that was marked by a difference in mathematical power. Certain kinds—Black, poor, and Hispanic—were seen as the kind of people who would disproportionately not acquire or did not already have the inherent power of mathematics, deemed necessary to live a productive kind of life.

In a similar style of comparative thought that produced standards of identity and difference in math reforms for equality, the mode of representing the mathematically literate as equivalent with progress and growth classified the mathematically illiterate as the cultural expression of difference and inequality. Mathematical illiteracy emerged as a social problem and provided new ways to see and think about the concern of organizing mathematics for all. The cultural organization of *all* seemed to be tethered to a concern that "changing demographics have raised the stakes for all Americans. Never before have we been forced to provide true equality in opportunity to learn" (NRC, 1989, p. 19). The standard notions of equality examined here through the equal sign were organized within a cultural imperative to manage a "true equality" wherein the mathematically illiterate was differentiated as a kind of person who lacked the power to solve the "real-world" problem of inequality.

Reforming People and the Limits of Standardizing Equality

Largely in response to *A Nation at Risk* (NCEE, 1983) and *An Agenda for Action: Recommendations for School Mathematics in the 1980s* (NCTM, 1980), the NCTM standards emerged as an attempt to overhaul math education to improve the economic, technological, and social lives of the nation and its people. This chapter has explored how the standards embodied a cultural set of principles that produced new ways seeing, thinking about, and acting upon children that were not available in the previous reforms. Specifically, it examined how the imperative to reform *math for all* worked to organize children as mathematical kinds of people and problem solvers. The continued analysis of the use of the equal sign in math problems highlighted how the logic of equality embedded in the commonsense that "everybody counts" embodied cultural standards about who counted as everybody.

As a shift from the previous back-to-basics and new math reforms, the classifications and distinctions that defined what it meant to be part of *math for all* during the standards-based reforms changed. With math education tied to a child's value as a person, the standards were rationalized as a way to create equal opportunities to benefit from math and to live a productive kind of life. Living as a mathematically literate and powerful—confident, expressive, and meta-aware—kind of problem solver became a new way of thinking about who the child is and should become through school mathematics. This kind of person appeared to use his or her individual powers "to attain the mature and informed judgment needed to secure gainful employment, and to manage their own lives, thereby serving not only their own interests but also the progress of society itself" (National Commission on Excellence in Education (NCEE), 1983, p. 9).

Still carrying notions of citizenship and individuality that were embedded in the previous reform efforts, the standards curriculum rearticulated how math could and should be for all through the expectation that everybody had inherent powers to use the language and reason of mathematics to solve the problems of everyday life. Related to the concern about making mathematics accessible and equally beneficial for all, it was through the use of mathematical powers that children were thought to have a greater sense of control over their present and future life choices, chances, and opportunities.

The discussion here highlighted how the practices of math pedagogy intersected with the principles of identity and equivalence expressed in the languages of cognitive and social psychologies. This illuminated the presumption that all people are similar in their mathematical powers and abilities as problem-solving kinds of people and showed how the standard problems of school math carried cultural norms organizing a person who *could see, think, and speak mathematically* about the everyday world. At stake in curriculum reform was the possibility of making all children into the kinds of people who interpreted the world through a power ascribed to math as a way of communicating about and solving problems in the world.

In ordering this kind of person, new ways to identify people who did not count as part of everybody emerged that were different from the previous reforms. Through the efforts to manage in/equality, children were classified as kinds of people who were either mathematically powerful or mathematically illiterate. Not "regardless of race or class or economic status" (ibid) but with regard to them as cultural expressions of a presumably deficient difference, some children appeared to lack the powers to confidently formulate, communicate, and rethink the mathematical modes of thought that organized the problems of everyday life.

What emerged during this time was an image of the child as a math person—or not—that has become part of the narrative of how math education should be reorganized for all. This mathematical kind of person embodies limits in ways of reasoning about equality through the reinscription of cultural notions of difference as the inequality that reform is to solve. Seen through presumably natural categories and distinctions, some children will always be included; others will never live up to the standard that defines the *all*.

References

Chavez, A. & Widmer, C.C. (1982). Math anxiety: Elementary teachers speak for themselves. *Educational Leadership, 39*(5), 387–388.

Cockcroft, W.H. (1982). *Mathematics counts.* London: Her Majesty's Stationary Office.

Johnson, S. (Ed.). (1983). *The world book of math power.* Chicago, IL: World Book Encyclopedia, Inc.

National Commission on Excellence in Education (NCEE). (1983). *A nation at risk: The imperative for educational reform: A report to the nation and the secretary of education, United States department of education.* Washington, DC: The Commission: [Supt. of Docs., U.S. G.P.O distributor].

National Council of Teachers of Mathematics (NCTM). (1989). *Curriculum and evaluation standards for school mathematics.* Reston, VA: NCTM.

National Research Council (NRC) Committee on the Mathematical Sciences in the Year 2000. (1989). *Everybody counts: A report to the nation on the future of mathematics education.* Washington, DC: National Academy Press.

Organization of Economic Cooperation and Development (OECD). (2009). *PISA 2009 assessment framework: Key competencies in reading, mathematics, and science.* Washington, DC: Author.

Paulos, J.A. (1988). *Innumeracy: Mathematical illiteracy and its consequences.* New York, NY: Hill and Wang.

Pimm, D. (1987). *Speaking mathematically: Communication in mathematics classrooms.* London: Routledge & Kegan Paul.

Schoenfeld, A.H. (1985). *Mathematical problem solving.* Orlando, FL: Academic Press, Inc.

Schoenfeld, A.H. (Ed.). (1987). *Cognitive science and mathematics education*. Hillsdale, NJ: Lawrence Erlbaum Associates.

Silver, E.A. (1987). Foundations of cognitive theory and research for mathematics problem-solving. In Schoenfeld, A.H. (Ed.), *Cognitive science and mathematics education*. Hillsdale, NJ: Lawrence Erlbaum Associates. pp. 33–60.

Skemp, R. (1971). *The psychology of learning math*. New York, NY: Penguin Books.

Steen, L.A. (1990). Numeracy. *Daedalus, 119*(2), 211–231.

Stodolsky, S.S. (1985). Telling math: Origins of math aversion and anxiety. *Educational Psychologist, 20*(3), 125–133.

7 The Alchemy of School Subjects and the Im/Possibilities of Reform

After awhile making a proof is like making a calculation.
There are certain things you automatically do.
You move with closed eyes, clenched eyes
unseeing eyes, no eyes.
You move, sometimes with no brain.
After awhile crossing the implication sign is like crossing the equal sign.
After awhile a proof is collapsed to a point.

(Deutsche Cohen, 2007)

This book has crossed the equal sign, tracing its use in the elementary math curriculum in the United States from the middle of the 20th century to the present. This historical crossing has considered how the equal sign carries cultural and social ways of reasoning about notions of in/equality. It has opened up a space to examine how the ways of seeing children as mathematical kinds of citizens, individuals, and people are not natural or neutral but political and cultural. Looking at the equal sign as it intersects with a social logic of in/equality has provided a way to see how rules about equivalence, identity, and difference order notions of equality and the child that is to learn it as part of the cultural expression of *all*.

The final chapter will continue to cross the equal sign with discussions of the paradox of remaking *math for all*. It first gives attention to how the norms of identity that are to include children in *math for all* have historically shifted from ideals of citizenship to distinctions of individuality to standards of personhood in ways that continually (re)define the terms of inclusion and exclusion. In doing so, it points to how the images of children as mathematical kinds have produced notions of difference as the problem of inequality. Considering the historically shifting identities and differences, this chapter questions contemporary reforms that take school subjects as the locus of change to interrogate the limits of situating the problem and solution to educational change in children.

By looking back across the previous chapters, this chapter provokes how the logic about making equality continuously reinserts differences in children as the problem and makes certain moves toward inclusion, equity, and access seem impossible. Yet in that impossibility is the possibility to rethink change in ways that do not place the problem in children. Through uncovering the historical contingencies in how children have been made to be seen as part of *math for all,* or not, this chapter and the book in general aim to explore the limits of reform embodied in the alchemy of school subjects and open alternative possibilities for change that do not rely upon remaking children.

Highlighting the politics of inclusion and exclusion at the site of elementary math education reform in the United States has been to understand the cultural and historical production of boundaries that characterize the child as a math learner. Yet the book is about more than mathematics. This chapter involves a discussion of how this research contributes to the study of schooling and curriculum, particularly in relation to issues of difference, diversity, inclusion, and equality. In that, it points to how the conditions of inclusion and exclusion are contingent and have been drawn along economic, gendered, racialized, and linguistic dimensions of culture in ways that cut across school subjects.

Although situated in discussions of current and historical math reforms that take school subjects as the origin of change, this chapter interrogates the emphasis on teachers as content experts across all subject areas and research that aims to prescribe best practices. In this way, it moves beyond the case of math to broadly question the limits of these reforms that rest upon fixed ways of knowing and (re)making children, locating the problem of research, and thinking about educational change.

"There Are Certain Things You Automatically Do": Remaking the Child and Difference Through the Alchemy of the Curriculum

The book began by signaling how school mathematics is not only about learning mathematics. This was approached by examining the double meaning of school subjects—that is, the content and how that content organizes pedagogical rules for who children are and should be—through the "alchemy" of school mathematics (Popkewitz, 2004). Each chapter has looked at how the fact of the equal sign comes into the curriculum as particular images, words, problems, and experiences for children through the language and tools of educational psychologies that make visible the cultural and political dimensions of math in schools. The alchemy—as a process of translation that makes mathematics into math education—has brought into focus the psychological models that organize children and their learning as the conditions by which they might be considered part of *math for all.*

Beyond a "pure" mathematics, the exploration of the curriculum as both the content *and* the process that orders how children should think, see, and act in the world has highlighted how the curriculum makes, or fabricates, children to have certain qualities. This ordering became visible through the tools and language of psychology as it assembled mathematics with social and cognitive concepts, like motivation, creativity, independence, and metacognition—in ways that have tied children's math learning to who children are and should be. Whereas these norms provide a way to see and describe children and the perceived problems in school math, they are not natural ways of thinking or talking about children. The discussion throughout the chapters has attended to how these classifications "make up" children as certain kinds of people (Hacking, 2002). This has been evident in how the psychological principles that translate math into the cultural spaces of the school organize and produce the image of children as certain kinds of citizens (Chapter 4), individuals (Chapter 5), and people (Chapter 6).

Highlighting how the norms of identity that aim to include all children into *math for all* have shifted from ideals of citizenship to distinctions of individuality to standards of personhood, the chapters pointed to historical contingencies that continuously redefine the terms of inclusion. With attention to a "double gesture" in how the production of identities also generates notions of differences (Popkewitz, 2008), the chapters also unveiled how the shifting rules for inclusion further distinguish who is not considered part of *math for all.*

The following sections will attend to the changing relationship between children's mathematics and the constantly shifting ways to classify and divide children. This historical overview preempts the discussion about the limits and im/possibilities for reforms aimed at (re)visioning society and remaking people within hopes for equality, inclusion, and justice. Whereas the argument continues to draw upon the math curriculum as the particular site of provocation, a more general discussion emerges to challenge the common sense of how schooling is thought about, its subjects studied, and reforms organized.

The historically shifting landscape of pedagogical practices revisited in the following sections signals implications for educational and curriculum studies. It provokes questions beyond the discussion of mathematics about how schooling has historically been positioned to evoke social change by remaking people in ways that continuously assume notions of unity, progress, and inclusion that simultaneously produce exclusions.

"After Awhile": Something New Will Emerge Again

In the middle of the 20th century, elementary school mathematics became part of an inclusion and equity imperative given in the emergence of *math for all*. Since that time, children's math learning has intersected with the cultural norms embedded in social and cognitive psychologies that organize children into the kinds of people who were presumed to make society more equitable, just, and democratic. As math education has been tied to social questions of children's participation as citizens, differentiation as individuals, and matters of personhood, its reforms have played a part in making the kinds of people who were to be seen as "good," "right," and "productive" in relation to notions of social progress, inclusion, and equality.

Crossing the equal sign across three major periods of elementary math reform, it has become evident how fabricating children as certain kinds of people is embedded in the logic of in/equality that circulates in the *math for all* reforms to produce equivalences and differences among children. This reading of the curriculum has offered a way to see how the reasoning about in/equality inscribed in the equal sign is given meaning and conditioned by historically shifting social, cultural, and political phenomena.

Whereas it has become common sense to think that all children can and should learn math, the inclusive imperative of school math has been expressed in different ways over time.

During the new math reforms after World War II in the United States, it became reasonable to think that all students can learn math *if* they can learn to think rationally, intelligently, and creatively through math's given structure (Chapter 4). The back-to-basics reforms circulated a way of thinking about how each child could learn math *when* they have a natural motivation and interest to learn it (Chapter 5). The reason of inclusion into *math for all* changed during the standards-based reforms to a logic about how every person could learn math *because* math is a logical way of seeing and communicating in the world to solve the problems of everyday life (Chapter 6). These truth statements have organized the reason of equality and served as conditions of inclusion in *math for all* reforms. Across three periods of reform, this reason has shifted and reinscribed cultural principles of identity and difference that have reshaped who will be included in the all and who will not.

As *math for all* emerged as a historical event in the post-World War II United States, notions of "intelligent citizenship" (NCTM, 1945) and "mathematical creativity" (Piaget, 1950) became part of the rationality that all children could and should learn mathematics. Examined in Chapter 4, the new math reforms were about more than mathematics in how elementary school math was tethered to notions of national progress, collective belonging, and remaking democracy. In this movement of *math for all* toward planning the future of the nation, children were imagined as citizens who could be developed as intelligent, independent, and creative through school math.

The intelligence and creativity mathematics was thought to produce became conditions by which it would seem possible for all children to be part of *math for all*. This was visible in how the use of the equal sign intersected with cognitive psychologies that assumed children could master the math content, given its common and universal structures. Through the psychological lens of developing modes of creative, abstract, and independent thought, all children were assumed to follow a common trajectory of learning. In this, moving from concrete to abstract thought was seen as a way to develop mathematical understanding and the psychological traits of a mathematical form of creativity—expressed through notions of flexibility, abstract reasoning, and independent thinking.

These cultural principles were embodied in the practices of learning math and an image of children as mathematically intelligent and creative citizens was inscribed in the practices of the new math curriculum. In the classroom, mathematical modes of creativity were to be internalized and would become visible in how children used materials to solve expressions such as $2 + 2 = \square$. As "a way of arranging knowns and unknowns in equations so that the unknowns are made knowable" (Bruner, Goodnow, & Austin, 1956, p. 7), this method of using concrete things to represent abstraction was more than a way of working with materials to produce understanding. It carried cultural norms that organized the child as able to think abstractly, precisely, and flexibly beyond the immediacy of physical objects—as the expression of mathematical creativity and intelligent citizenship.

Given as a new way to see children, "mathematical creativity" signified the space of inclusion wherein all children could be "intelligent citizens" who use math as a "tool of modern life" (CEEB, 1959). In this hope was also the production of fear about a child who was not seen as mathematically intelligent and represented a threat to living the modern life. Emerging as a new kind of social and pedagogical problem, "deprived" and "disadvantaged" children appeared in both the new math curriculum and broader educational reform policies.

The distinctions that classified a "deprived" or "disadvantaged" child as different were recognizable in relation to what defined normative modes of mathematical creativity. That is, "the mental style of the socially and economically disadvantaged learners resembles the mental style of *one type* of highly creative persons. Our schools should provide for the development of these unique, untapped national sources of creativity" (Riessman, 1964, p. 231). This development relied upon seeing a different kind of creativity that characterized the presumably disadvantaged child as a singled-minded, slow, spatially oriented, and concrete thinker. These traits and the different kind of creativity they assumed were compared to desired modes of thinking and living.

The child seen as belonging to the disadvantaged, impoverished, lower-class culture was presumed to develop a "different kind of mind" (Havighurst, 1964, p. 211). This difference was expressed as "certain personal deficits" and were made visible as a restricted use of language, inferior auditory discrimination, inferior visual discrimination, inferior judgment concerning time, number, and other basic concepts (ibid, p. 214). This child had apparently not developed the ability to listen, see, pay attention, or talk about the world in particular ways.

Embodying the hope of making all children into intelligent and mathematically creative kinds of citizens, the differences were given as a way to understand and reason about problems of inequality in education. The problem was placed in the child's social and economic conditions—given as a geographical distinction. It seemed that "the socially disadvantaged children tend to come from families that are poor and recent immigrants to the big cities" (ibid, p. 215). This geographical categorization became a racial and ethnic way of classifying the "deprived" or "disadvantaged" math student. This classification encoded cultural significations that differentiated poor, Black, Hispanic, and Puerto Rican students as the kind of children who were seen at a disadvantage to living the modern life as a rational, creative, independent, and open-minded citizen through learning mathematics.

In the period of back-to-basics reform following new mathematics, ideals of creativity merged and reassembled with psychologies based upon ideals of each individual child's differences and discovery-based learning. With that encounter came new psychological traits that traveled with the equal sign and re-signified children as mathematically motivated kinds of individuals. The shifting norms of inclusion were expressed no longer as mathematical forms of intelligent citizenship through creative and abstract thinking but as a personal motivation to learn mathematics in one's own self-interest.

In this movement from new math to back to basics, there remained a continuity in the focus on the child as the way to plan the future through the inscription of traits deemed desirable for that future. This is the alchemy, and it is not only in mathematics. This is evident in current expressions of the need for educational reforms focused on the intersections of STEM education where "preparing the next generation of STEM innovators" becomes a project of "identifying and developing our nation's human capital" (National Science Foundation, 2010). The common sense follows the logic that the coupling of STEM will create a formula for national stability and survival via the fabrication of "innovators" imbued with a sense of value as certain kinds of people. Within the belief in making investment in "human capital," the reforms are rationalized as a way for society to build abilities and skills in areas considered to be useful knowledge for modernizing democracy.

The limits of this proposal for change become evident in how taken-for-granted ways of thinking about (re)visioning society and people continually identify some as the hope for development, survival, and inclusion while reinserting difference as the problem of reform, fear of decay, and reason for inequality. The possibilities and limits—bound by the historical changes that re-signify what constitutes progress—emerge together as the norms that define the rules of inclusion and rationalize the rules of exclusion in ways that make a modern and democratic life seem im/possible.

"You Move with Closed Eyes, Clenched Eyes": Back to the Basics of Motivating Each Child

Studying the alchemy of the curriculum during the back-to-basics movement in the 1970s in the United States, it became apparent how the logic shifted from all children can learn math *if they think creatively* to all children can learn math *when they are motivated and interested in it*. The curriculum was reorganized as a way to develop what was seen as a most basic form of motivation. Chapter 5 highlighted how, in part, this change was tied to beliefs that math's structure was too abstract and not linked to children's natural interests or motivations. If school math was organized around the abilities, interests, and skills of children, rather than with abstract mathematical structures, then it seemed reasonable to think that every child could be motivated to excel in basic mathematics.

With motivation and its psychological traits given as the condition for equal opportunities and the right to a quality education, inclusion was redefined. About more than basic math skills, the psychological norms of self-discovery, -interest, -control, and -awareness assembled with the math curriculum to regulate who each child was and should be as a mathematically motivated individual.

In the previous reform, the conditions for inclusion were contingent upon a psychological rule that all children had a common way of developing along a trajectory of thought. In getting back to basics, this rule intersected with the belief that children learn, think, and behave differently as individuals. These differences, expressed as abilities in math learning, became the focus of the psychologies that organized the math curriculum during that period of reform.

The pedagogical problem, then, was not one of providing the opportunity for all children to think creatively; it was to provide optimal opportunities for each child to be motivated to discover one's self and one's own mathematical abilities and interests. Enhancing the so-called natural motivation of each individual child provided a psychological lens through which to create an "equality of optimum motivation" (Nicholls, 1979). With this aim, the logic of individualizing instruction rested on the promise of motivation. Although all children were different in their interests, needs, and strengths, they all seemed to have a natural motivation, intuition, and ability for learning basic math skills.

This way of reasoning about math learning as a matter of differences in personality, interests, and intuitions assembled with the practices of problem-solving in the curriculum and reimagined the child as a mathematically motivated kind of individual. Mathematics in this basic form was to be organized around a "natural motivation" to bring the study of math closer to each child and allow for the discovery of math as an interesting and useful system of thought (Kline, 1973, p. 149).

In the curriculum, this focus on the child's interests and motivations translated into a "Child-created Mathematics" (Cochran, Barson, & Davis, 1970). Recall the discussion from Chapter 5 about Leslie and the problem of choosing numbers and creating a graph to explore the expression $\Box - 3 = \Delta$. This problem was to provide the opportunity to discover the equivalence between $\Box - 3$ and Δ on either side of the equal sign. A basic math skill was given as the practice of representing shapes as numbers to create an expression of equality. But it was not just about math. This problem intersected with cultural norms of children's interests, abilities, and choice making as a motivation to learn math.

Equally motivating all students based on a view of their individual abilities relied upon a pedagogy of mathematics that shifted emphasis from a structured and external knowledge to be mastered to one that was to be discovered and internally negotiated through each child's own sense of self-direction, control, and interests. In this shift, the traits ascribed to children were determined primarily through a psychology that sought to understand the individual differences among children who were doing well in math and those who were not.

In the larger scope of such studies, particular attention was given to understanding the distinctions in precocity drawn along gendered lines (Astin, 1974). With mathematically precocious girls characterized as "naturally" more sympathetic, tender, conscientious, and shy, it seemed quite clear that they were not like their male counterparts. It appeared to be the exception, rather than the rule, that girls could excel in math. In this classification, it seemed that a girl's exceptionality was based upon a comparison with the normal attributes of the precocious boy whose achievement in mathematics was apparently influenced by factors such as sociability, self-acceptance, well-being, responsibility, and self-control (Weiss, Haier, & Keating, 1974). The qualities given to characterize a girl's mathematically precocious sense of self were taken to explain why they were not motivated to achieve to the level of the boys.

Gendered norms and distinctions that further categorized the child who could or could not learn basic math also intersected with racial classifications. The child who seemed to have problems with math was still identified as "disadvantaged," as in the previous new math curriculum (Kaplan, 1970). Yet this disadvantage was given finer distinction in relation to the norms of self-control, -discipline, and -discovery.

The cultural signification of the mathematically motivated self-produced a pathology represented as the "mathematically disabled," "inner-city," child with "problems of motivation" who did not seem motivated to learn math became visible as a problem of difference in the curriculum (Keiffer & Greenholz, 1970). This "inner city" geographical distinction was resignified as a cultural distinction, assigning trouble with learning math to students who "are mainly Black, Spanish, and other minorities. They come from inner-city areas, and, in fact, thereby have developed negative attitudes toward mathematics" (Williams, 1974, p. 650). As the classification of "inner city" moved out of the city, the disabled and unmotivated math learner became visible as the "minority student." It appeared to be possible to explain mathematical forms of disability with economic, gendered, and racial categorizations.

Motivation, in its varying expressions explored here, emerged as a commonsense way of reasoning about how to reach all children through reorganizing math curriculum and pedagogy. This link between a seemingly natural motivation to learn math and individual achievement was part of the alchemy that historically produced an image of children who are not interested in and cannot control their own success in mathematics. Certain children could be seen to have the choice, opportunity, and potential to do well in mathematics yet lack the motivation or interest to do so. With the differences historically reinscribed as a "problem of motivation," the problem of pedagogy became one of remaking the child's self-image, controlling interest, and organizing discovery as a way to manage inequality.

The comparisons deemed useful in identifying and correcting social wrongs through educational reforms have been evident since the early decades of the 20th century as part of the politics of schooling. The opening up of American schools for children of immigrants and racial groups carried with it a way of thinking about what are now called changing demographics that produced distinctions of a backward, uncivilized child who appeared to not succeed in schools or society. These classifications have shifted across time and space and work through the curriculum to make children into certain kinds of people and define the boundaries of in/exclusion.

The double gesture of inclusion and exclusion was further explored in Chapter 6 as the continued promise *for all* shifted the historical ways of thinking about the relationship between children's education and their identities. In this continuity, there were changes as further nuance was given to determine who could be included. Through the alchemy, it became evident how as notions of motivation and discovery learning moved into the next period of reform, they intersected with a psychology of problem-solving—to give all children confidence in their seemingly natural abilities to speak, act, and think as a mathematical kind of person.

This kind of person does not simply exist. It has been created over time as some "thing" to see in schools, act upon in the curriculum, and study in educational research. This shifting signification of kinds of people can be seen in discussions of how various models of curriculum carry the desired characteristics that that children are to internalize through learning science through a "scientific mind" (Kirchgasler, 2017) or art as the "developed" child (Martins, 2017), for example. Given different categorizations, each of these discussions highlights how the form and content of the curriculum is more than the disciplinary knowledge it references in how it embodies cultural rules that (re) assign identities to children through psychological theories about the nature of the "normal" child that produce classifications of difference.

"You Move, Sometimes with No Brain": A Standard Approach to Making Everybody Count

As the psychological principles of learning math shifted from mastery learning to an assumption that children would discover their own mathematics to an organization of problem-solving relying upon an inherent logic in mathematics and social communication, the ordering of motivation shifted and became reassembled as a mode of organizing confidence as a problem solver. Emerging during the standards-based movement of the 1980s as a new way to organize math for all children, confidence became the lens through which to evaluate problem-solving and its expression of people's presumed "mathematical power" (National Research Council [NRC], 1989). This confidence was to be built through the use of math as a seemingly universal language to communicate with others about everyday problems in the world as a form of what was given as "mathematical literacy" (ibid).

During the reorganization of children's math education according to standards, the association between language and mathematical thinking framed the expectations by which children were to speak and think mathematically. Given attention in Chapter 6, a way of reasoning about how *everybody counts* emerged and implied a cultural logic of an equal benefit to be achieved through a notion of mathematical power. Presumed as something common to all people, mathematical thinking and its given powers were seen as a natural capability—something all people do as an intuitive process. Learning to think and express oneself mathematically was taken as a developmental norm and gave validity to the idea that all children could "grow in their ability to communicate mathematically and use higher-level thinking processes" (National Council of Teachers of Mathematics [NCTM], 1989, p. 23). Communication in general and the use of a mathematical kind of language presumably nurtured a growth into the kind of person who could confidently speak about mathematical thought processes.

As problem-solving practices intersected with the rules of cognitive and social psychologies during the standards movement, a new way of seeing children as mathematical kinds of people emerged. Identifying children as social beings who learn in various ways, yet always through interactions with others and the world, the rules of psychology assembled with the logic of mathematics to produce who the mathematical problem solver was and should be through common expressions of mathematical power. In solving problems similar to the expressions of $14 - 5 = \square$, as discussed in Chapter 6, children were to model each problem as a mode of translation so they could see and discuss how they were different to see their similar underlying structure. This modeling, translating, and seeing the problem was to establish a common way for children to visualize and communicate about the problems.

These standard expectations entangled with research that emerged to understand *Mathematical Problem Solving* (Schoenfeld, 1985) as a "spectrum of behaviors that comprise people's mathematical thinking," (p. 5). These behaviors were bound to the use of language and symbolic expressions to represent internal thoughts to others. This required, it seemed, the ability of the person with math problem-solving abilities to self-monitor and be aware of his or her own thought process. Given as traits of the "good problem solver," meta-awareness and self-regulation organized how to see a child's ability to communicate mathematics in what would appear to be an organized and logical way (Schoenfeld, 1987). Ordered by psychological principles identifying effective problem-solving strategies, children confident in their mathematics were seen as people who could make sense of, plan, communicate, rethink, and represent a well-organized, logical, and justifiable mathematical thought process.

Marking a shift in emphasis from all children as self-motivated individuals, the standards organized and maintained that all children could learn math as strategic and confident problem solvers. Motivation was still an important psychological factor in learning math, but it became reassembled as confidence according to the logic that everybody could reason about and communicate math to others. This confidence framed a standard approach to all students learning math because it was thought that the confidence was to be built from the validity granted to math itself. In fact, the standards maintained that the assessment of mathematical power "should examine students' disposition toward mathematics, in particular their confidence in doing mathematics and the extent to which they value mathematics" (NCTM, 1989, p. 205). Taken as psychological dispositions of people who effectively solved problems, confidence and an intentional appreciation of math were associated with particular learning behaviors and, therefore, possible to evaluate. The shaping and reshaping of mathematical thought, then, was tied to a reorganization of the child's value system as well as his or her value as a person.

Within a movement toward math standards, mathematical power made it reasonable to think that everybody could be a problem-solving kind of person. These aims were instantiated in how children were to see, think, speak, represent, and communicate mathematically—as the expression of mathematical power. It seemed to make sense, then, that the standards were to "ensure that all students benefit equally from the opportunities provided by mathematics" (NCTM, 1989, p. 89). The reasoning about this equal benefit was embodied in the goals of the standards to provide experiences for all students to value math, be confident in one's abilities, become a mathematical problem solver, and communicate and reason mathematically (NCTM, 1989, pp. 5–6). These goals assumed the kind of person who appreciated math, respected its certainty, understood and spoke its language, and saw through its perspective to confidently solve the problems of everyday life.

Embedded in the equivalence given to children as equally capable of possessing and demonstrating mathematical power was a comparative frame with which to see children who apparently could not draw upon a mathematical power to benefit from all that mathematics was to provide. This was examined in Chapter 5 through discussions of the historical emergence of mathematical illiteracy as a problem of inequality and difference (NRC, 1989). The mathematically illiterate was fabricated as a kind of person who apparently could not appreciate the sense of certainty a mathematical perspective would lend to one's approach to living and problem-solving in the social world. To use another language, they were seemingly an insecure and passive group who left life to chance, could not assess benefit and risk, and were otherwise unplanned, unpredictable, and unreliable (Paulos, 1988). Not merely about math, the distinction appeared in relation to the kind of person who was desired—confident, meta-cognitively aware, and expressive of logical thoughts.

As a problem in the pedagogy of mathematics, the differences that were produced to identify innumerate kinds of people as insecure, disengaged, ignorant, and unreflective were conceived within and emerged alongside the psychological traits that produced a mathematically powerful kind of person. Seen as less productive and overall lacking the power to act in intentional, rational, and organized ways, this classification of difference as the "mathematically illiterate" was a shift from earlier reforms. Given a new name, the markers of difference that defined the problems of inequality were no longer articulated as "disadvantaged," "disabled," or "inner-city" kinds of people as in the previous reforms. Yet the shift to a "mathematically illiterate" kind of person carried the same comparative way of reasoning about difference and the problems it presumes.

Mathematical literacy—as the expression of a mathematical kind of language, power, and personhood—has embedded in it the ability to see, communicate, (re)think, and make sense of the problems that constitute the everyday. It also carries with it the limits and historical conditions whereby the possibility to include all children in schools does not make much sense and, in fact, seems impossible.

This im/possibility of inclusion captures the "partition of the sensible," which defines the conditions within which the norms characterizing who children are and should be through mathematics are produced to intern and enclose what is reasonable in the present (Rancière, 2010). Including all students and granting access to skills and knowledge for future success makes sense. This seems possible. Yet this commonsense way of reasoning has left the cultural politics and history of remaking education for societal change unquestioned.

"After Awhile Crossing the Implication Sign Is Like Crossing the Equal Sign": The Double Gesture of In/Equality, In/Justice, and In/Exclusion

Paradoxically the hope of making children into productive citizens, rational decision makers, and creative and independent thinkers who are self-aware, motivated, expressive, and logical carries a fear of the kinds of children who are not and cannot become that. This paradox—the insertion of difference as the reason of inequality in efforts toward equality and inclusion—has been explored throughout the book through the analytical theme of a "double gesture" (Popkewitz, 2008). The chapters work together to open up questions about the making of *math for all* and how this possibility is limited and enclosed by the fabrication of children as the kinds who represent the hope for equality and inclusion, distinguished from those characterized as different and the problem of inequality.

These traits have historically emerged as the conditions by which *all children* can learn math such that it seems possible to identify the kinds of children who cannot. This historical and contingent production of identities and differences, explored in each chapter, names some children as part of *the all*, whereas others represent the problem of inequalities in education. The distinction between the child who fits and the child to be fixed is drawn along cultural dimensions that presume the economically, racially, gendered, and linguistically diverse child as already excluded from the norm. Across the chapters, this was evident in how the fabrication of productive citizens; expressive, creative, independent individuals; and self-controlled, self-directed, confident kinds of people has also created distinctions expressed as mathematically disadvantaged, disabled, and illiterate.

The differences produced have been further distinguished and given new terminology—always within a representational and normative logic about identity. That is to say, the inscription and production of difference as a problem situated in "other" kinds of children is part of the logic that equality and inclusion require the identification of what constitutes the *all*. The attention given to this double gesture—of inclusion/exclusion, hope/fear, and in/equality—brings into focus how efforts to improve social conditions are worked on through principles that produce comparative modes of reasoning. This logic circulates and reinscribes the very notions of difference as the representation of an inequality they aim to upset.

Without accounting for how the curriculum continuously inscribes comparative modes of reasoning about who children are and should be through schooling, it is impossible to see how efforts to raise levels of achievement and reduce inequalities in education and society rely upon comparative measures. The statistics, comparative models, and psychological traits that assemble to create the picture of existing differences as deficiencies in particular kinds of students become rationalized as the reason for inequalities. The problem of change, then, is posed as the prescription of normative commitments to improve opportunities to learn content as a way to fix the kind of child that represents a gap in access and achievement.

As a comparative mode of reasoning, the simultaneous inclusion/exclusion is not natural or neutral. The inscriptions of identity and difference are not given to children to describe what is already there. A so-called disadvantaged or disabled child does not exist as such but is made up through a comparative lens in reference to the hoped-for kind of child. Treated here as historical classifications and divisions that continuously re-signify who does or does not belong to the inclusive spaces of schooling for all, the ways to describe children and account for their differences are open to question and change.

This attention to the double quality of school subjects—the content and how that content organizes who children are and should be—shifts the focus in contemporary discussions of curriculum and instruction about teachers as content experts and research prescribing best practices. The current emphasis on what children and teachers need to know in literacy, math, and science is central to reforms in schooling. But this takes the school subjects as neutral and situates the problem of change either in the child or the curriculum—with the teacher as an "expert" tasked with a prescription of "best practices."

This way of reasoning about how to see, think about, and act upon children as certain kinds preempts a seemingly rational conclusion: If only the "disadvantaged," "struggling," "low-achieving" children were more motivated, independent, and confident, then they would succeed. This logic situates the problem in the child and elides questions about the production of normative traits that have historically (re)shaped who the child should be through schooling: creative, rational, logical, motivated, self-aware, independent, confident, and expressive.

What this study of the curriculum points to, instead, is how those practices, curriculum models, and theories of children are open to question. If, as the book has shown, the identities and differences ascribed to children are not natural and historically shift according to cultural and political ways of reasoning about who children are and should become through schooling, then the comparative modes of reasoning embodied in the reforms are to become the problem of research and reform, not the children.

"After Awhile a Proof Is Collapsed to a Point": The Research of Curriculum and Its Reform

"If everything is dangerous, then we always have something to do" (Foucault, 1984, p. 343). This *something to do* has been captured in the book's disruption of the status quo through raising questions about the systems of reason that relate cultural and mathematical ways of reasoning about in/equality—not as stable facts but as historically contingent rules about organizing equivalence and difference. From this view, it has become possible to rethink how the very reorganization of curriculum, through its interactions with psychology as "truth," produces and maintains theories about children that operate to make them appear with certain qualities and traits.

As an approach to studying curriculum, reform, and the logic of schooling, the book has opened up a space to examine how the practices of classification—taken as neutral and natural elements of a school curriculum—embody limits in their reason. As historical and not necessarily natural, the taken-for-granted rules and practices that order pedagogical thought and action in mathematics education as explored here can be seen as unfixed. That is, insofar as the categories that identify and differentiate children are historically shifting, they are contingent and open to question. Whereas studies that propose moves toward inclusion, examine equity and achievement gaps, and acknowledge diversity are important, this step to the side has been to understand the historical ways of reasoning that have produced the now taken-as-granted identities and differences that comprise the gap, define diversity, and prescribe what counts as equity.

Importantly, "it is through locating changes in the rules of reason that we can think about change" (Popkewitz, 2001, p. 168). By identifying how the ways of thinking about who children are and should be through schooling continuously shift, then new ways of seeing, knowing, and acting can be made possible. Through an interrogation of the historical rules that organize children as particular kinds, it becomes possible to see that these have not always been available ways to see, talk about, or act upon children. Approaching the study of curriculum and education as a "cultural history of the present," the book has shown how the present categories that organize children's identities and distinctions have historically become "real" categories that make up who the child is and should be in ways that are not natural or neutral (Popkewitz, Franklin, & Pereyra, 2001). The common sense of the present has been and could be otherwise.

From the perspective of focusing on the rules of reason for in/equality, both the problem and the origin of change in schooling does not rest in children, teachers, psychologists, mathematicians, policy makers, parents, school administrators, or curriculum developers. No single historical actor is to be taken as the locus of reform. Change and reform are thus reimagined by examining how the historical conditions that organize the common sense of the present embody limits in their ways of reasoning about what constitutes progress, democracy, inclusion, diversity, access, and so on. By resituating the problem in the logic of representation and difference that operates through the curriculum, translated in the language of psychology, the solution is not one that can easily be prescribed or described as a way to "fix" those who stand as the reason for inequality. That way of rethinking the problems and solutions in educational reform does not offer prescribed solutions, only unplanned alternatives and a renaming of the problem.

Whereas this study takes elementary math curriculum reforms and research in the United States as the site of inquiry, it also speaks more broadly to how equality is instantiated as a cultural practice in the curriculum. In arguing that the equal sign is related to a theory of equality that is not merely mathematics itself, the book showed how the cultural norms of equality, identity, and difference give meaning to present reforms and instructional practices. The math curriculum has provided a site, like science, literacy, or art education, to explore how efforts to improve social conditions are worked on through comparative modes of reasoning that reinscribe notions of difference as the representation of an inequality they aim to upend. Unraveling the historical relations has worked to understand the political implications of reforms that produce differences as inequality, identifying some children as equals, whereas others are not.

More than about mathematics education, this book is about the possibilities that might open through rethinking how the subjects of schooling have been constituted through notions of identity and difference. It questions how the curriculum reforms are tangled up with ways of thinking about social in/equality, the politics of identity and difference, and in/exclusion. Whereas its analyses have addressed the paradoxes present in educational reforms and research concerned with equity, inclusion, and justice, this is not to say these are not important commitments in schooling. But when the modes of reasoning about the reforms aimed at including *all* are examined, the *all* is not inclusive—evident in the production of difference that simultaneously excludes.

Recognizing the simultaneous production of both identities and differences embodied in the cultural ways of reasoning about in/equality in curriculum reforms aimed at *all students* provokes how in the recurring calls for diversity and inclusion the "greatest danger is lapsing into the representations of a beautiful soul which says: we are different but we are not opposed" (Deleuze, 1994). It is this danger of representation that points to the conditions by which difference has been historically constructed in opposition to that which is seen as naturally beautiful, valuable, and appreciated. Even in recognizing and valuing difference, the difference is assumed and goes unquestioned in terms of how it is historically produced and produces political implications for in/exclusion.

In this project, then, is the hope that with an understanding of the past as conditions that shape the present, change in the form of rethinking what is known might be possible. With in/equality as both a problematic of the curriculum and a question open for consideration, alternative ways of reasoning about educational reform and children may emerge. This hope embodies a double gesture of fear—a fear that even within the well-intentioned efforts to (re)organize education as equal, equitable, and inclusive, the ways of thinking that rationalize the dividing practices embedded in the curriculum will continue to go unquestioned.

References

Astin. (1974). Sex differences in mathematical and scientific precocity. In Stanley, J.C., Keating, D.P., & Fox, L.H. (Eds.), *Mathematical talent: Discovery, description, and development*. Baltimore, MD: The Johns Hopkins University Press. pp. 70–86.

Bruner, J., Goodnow, J.J., & Austin, G.A. (1956). *A study of thinking*. New York, NY: John Wiley & Sons, Inc.

Cochran, B.S., Barson, A., & Davis, R.B. (1970). Child-created mathematics. *Arithmetic Teacher*, March, pp. 211–215.

College Entrance Examination Board. (1959). Program for college preparatory mathematics. In Bidwell, J.K. & Clason, R.G. (Eds.). (1970). *Readings in the history of mathematics education*. Washington, DC: National Council of Teachers of Mathematics. pp. 664–706.

Deleuze, G. (1994). *Difference and repetition* (P. Patton, Trans.). New York, NY: Colombia University Press.

Deutsche Cohen, M. (2007). *Crossing the equal sign*. Austin, TX: Plain View Press.

Foucault, M. (1984). *Foucault reader* (P. Rabinow, Ed.). New York, NY: Pantheon Books.

Foucault, M. (1988). *Politics, philosophy, culture: Interviews and other writings 1977–1984*. (L.D. Kritzman, Ed.). New York, NY: Routledge.

Hacking, I. (2002). Inaugural lecture: Chair of philosophy and history of scientific concepts at the College de France. *Economy and Society*, 31(1), 1–14.

Havighurst, R. (1964). Who are the socially disadvantaged? *Journal of Negro Education*, 33(3), 210–217.

Kaplan, J.D. (1970). An example of a mathematics instructional program for disadvantaged children. *Arithmetic Teacher*, April, pp. 332–334.

Keiffer, M. & Greenholz, S. (1970). Don't underestimate the inner city child. *Arithmetic Teacher*, November, pp. 587–595.

Kirchgasler, K. (2017). Scientific Americans: Historicizing the making of difference in early 20th-century U.S. science education. In Popkewitz, T., Diaz, J., &

Kirchgasler, C. (Eds.), *A political sociology of educational knowledge: Studies of exclusion and difference*. New York, NY: Routledge. pp. 87–102.

Kline, M. (1973). *Why Johnny can't add: The failure of new math*. New York, NY: St. Martin's Press.

Martins, C. (2017). From scribbles to details: The invention of stages of development in drawing and the government of the child. In Popkewitz, T., Diaz, J., & Kirchgasler, C. (Eds.), *A political sociology of educational knowledge: Studies of exclusion and difference*. New York, NY: Routledge. pp. 103–116.

National Council of Teachers of Mathematics. (1989). *Curriculum and evaluation standards for school mathematics*. Reston, VA: NCTM.

National Research Council (NRC) Committee on the Mathematical Sciences in the Year 2000. (1989). *Everybody counts: A report to the nation on the future of mathematics education*. Washington, DC: National Academy Press.

Nicholls, J.G. (1979). Quality and equality in intellectual development: The role of motivation in education. *American Psychologist, 34*(11), 1071–1084.

Paulos, J.A. (1988). *Innumeracy: Mathematical illiteracy and its consequences*. New York, NY: Hill and Wang.

Piaget, J. (1950). *The psychology of intelligence*. New York, NY: Harcourt, Brace & Co., Inc.

Popkewitz, T.S. (2001). The production of reason and power: Curriculum history and intellectual traditions. In Popkewitz, T., Franklin, B., & Pereyra, M. (Eds.), *Cultural history and education: Critical essays on knowledge and schooling*. New York, NY: RoutledgeFalmer. pp. 151–183.

Popkewitz, T.S. (2004). The alchemy of the mathematics curriculum: Inscriptions and the fabrication of the child. *American Educational Journal, 41*(4), 3–34.

Popkewitz, T.S. (2008). *Cosmpolitanism and the age of school reform: Science, education and making society by making the child*. New York, NY: Routledge.

Popkewitz, T.S., Franklin, B., & Pereyra, M. (2001). *Cultural history and education: Critical essays on knowledge and schooling*. New York, NY: RoutledgeFalmer.

Rancière, J. (2010). On ignorant schoolmasters. In Bingham, C. & Biesta, G. (Eds.), *Jacques Rancière: Education, truth, emancipation*. New York, NY: Continuum. pp. 1–24.

Riessman, F. (1964). The overlooked positives of disadvantaged groups. *The Journal of Negro Education, 33*(3), 225–231.

Schoenfeld, A.H. (1985). *Mathematical problem solving*. Orlando, FL: Academic Press, Inc.

Weiss, D., Haier, R.J., & Keating, D.P. (1974). Personality characteristics of mathematically precocious boys. In Stanley, J.C., Keating, D.P., & Fox, L.H. (Eds.), *Mathematical talent: Discovery, description, and development*. Baltimore, MD: The Johns Hopkins University Press. pp. 126–139.

Williams, E.R. (1974). Mathematics for the "disadvantaged." *The American Mathematical Monthly, 81*(6), 648–659.

Index